Human Systems Integration for Mining Automation

Human Systems Integration for Mining Automation is the professional's guide to understanding the issues, approaches, and pitfalls associated with mining automation from a human perspective. This book delves into a timely and fast-developing issue in mining and the wider minerals industry - the design and deployment of automation.

The book approaches this from a "Human Systems Integration" standpoint in which the technical and human-related aspects are jointly considered as part of an integrated, automated mining system. This accessible and readable title offers a wider Human Systems Integration framework that can be applied to mining projects. It is based on an established framework that has been developed and used successfully in other work. The framework is backed up with information obtained from mines in Australia, the USA, Canada, Sweden, and Chile and original equipment manufacturers such as Caterpillar, Komatsu, Sandvik and Epiroc. Every reader of this book will recognise the essential benefits of human systems integration for mining automation.

This book will be an ideal read for industry professionals including systems engineers, safety engineers, mining engineers, human factors engineers, and engineers working on developing and deploying automation in mining and related industries including rail, road transport, and process control. It will also be of interest to students, researchers, and academics in related fields.

Human Systems Integration for Mining Automation

Robin Burgess-Limerick, Tim Horberry,
Danellie Lynas, Andrew Hill, and Joel M Haight

CRC Press
Taylor & Francis Group
Boca Raton London New York

CRC Press is an imprint of the
Taylor & Francis Group, an **informa** business

Designed cover image: image credited to Evgeny_V [ShutterStock ID:1193990185]

First edition published 2025
by CRC Press
2385 NW Executive Center Drive, Suite 320, Boca Raton FL 33431

and by CRC Press
4 Park Square, Milton Park, Abingdon, Oxon, OX14 4RN

CRC Press is an imprint of Taylor & Francis Group, LLC

ISBN: 978-1-032-44709-4 (hbk)
ISBN: 978-1-032-46278-3 (pbk)
ISBN: 978-1-003-38088-7 (ebk)

DOI: 10.1201/9781003380887

Typeset in Times
by Deanta Global Publishing Services, Chennai, India

Contents

Acknowledgements

We acknowledge the generous sharing of knowledge and experience by the many people within mining companies and equipment manufacturers who participated in interviews, focus groups, and informal conversations. We are also grateful for the cooperation of the sites that hosted visits and contributed to case studies. The research was supported by funding from ACARP and the National Institute for Occupational Safety and Health (USA). We thank Prof Maureen Hassall for writing Chapter 6.

About the Authors

Professor Robin Burgess-Limerick is a Professorial Research Fellow in the Minerals Industry Safety and Health Centre at the University of Queensland in Australia. Robin is a past-President and Fellow of the Human Factors and Ergonomics Society of Australia. He is the second author of "Human-Centered Design for Mining Equipment and New Technology" (2018) and "Human Factors for the Design, Operation, and Maintenance of Mining Equipment" (2011).

Professor Tim Horberry is Professor of Human Centred Safe Design at the University of Queensland in Australia. He is the first author of "Human-Centered Design for Mining Equipment and New Technology" (2018) and "Human Factors for the Design, Operation, and Maintenance of Mining Equipment" (2011).

Dr Danellie Lynas is a Senior Research Fellow within the Minerals Industry Safety and Health Centre, Sustainable Minerals Institute at the University of Queensland, Australia. She is a Certified Professional Ergonomist and has previously conducted research in the area of mining automation.

Associate Professor Andrew Hill is Principal Research Fellow within the Minerals Industry Safety and Health Centre, Sustainable Minerals Institute, The University of Queensland. He has a PhD in Cognitive Psychology, and has conducted research in applied cognitive psychology across diverse industries.

Professor Joel Haight is Professor of Industrial Engineering, Swanson School of Engineering at University of Pittsburgh, USA. He has a PhD in Industrial and Systems Engineering from Auburn University and has spent 21 years as researcher and professor, 4 years as research manager in U.S. Federal government, and 18 years as an engineer and manager in the oil industry.

Contributors

Maureen Hassall, Guest author, Professor and Director, Minerals Industry Safety and Health Centre, Sustainable Minerals Institute, The University of Queensland, Australia

1 Human Aspects of Mining Automation

1.1 INTRODUCTION

The overall impact of increased mining equipment automation is likely to be improved safety and health. However, introducing autonomous components creates new failure modes. While guidelines for the implementation of autonomous mining equipment exist, these documents pay insufficient attention to the integration of humans and technology. This book puts a spotlight on the human aspects of automation in mining by adopting a human systems integration framework. Human systems integration refers to a set of systems engineering processes originally developed by the Defence industry to ensure that human-related issues are adequately considered during system planning, design, and development (INCOSE, 2015).

Human Readiness Levels provide a metric that shows the preparedness of a system to be used by human operators and maintainers (Handley, 2023). They are designed to complement Technology Readiness Levels that have been widely accepted as a means to assess the maturity of a technology before and after it is integrated into a developing system. The Human Factors & Ergonomics Society (HFES)/American National Standards Institute (ANSI) Human Readiness Level standard provides a way to evaluate, track, and show the readiness of a system for human use (HFES/ANSI, 2021). The standard provides supporting questions for each level to ensure human-related considerations are appropriately addressed and is an objective, structured method for assessing a system's readiness for human use. The goal of the Human Readiness Levels scale is therefore to provide a metric that indicates the state of the system with respect to the integration of humans and technology. A summary of the nine Human Readiness Levels is shown in Table 1.1.

1.2 IMPROVING HEALTH AND SAFETY THROUGH THE AUTOMATION OF MINING EQUIPMENT

While almost certainly not the first to do so, Corke et al. (1998) highlighted the potential for "advanced robotics systems" to reduce the exposure of miners to fatality and injury risks, as well as health hazards such as noise, vibration, dust, and diesel particulates – and this assertion has been frequently repeated as a motivation for automating mining equipment (e.g., Corke et al., 2008; Fisher & Schnittger, 2012;

DOI: 10.1201/9781003380887-1

TABLE 1.1
The nine-level "Human Readiness Levels" scale

1	Relevant human capabilities, limitations, and basic human performance issues and risks identified
2	Human-focused concept of operations defined, and human performance design principles established
3	Analyses of human operational, environmental, functional, cognitive, and physical needs completed, based on proof of concept
4	Modelling, part-task testing, and trade studies of user interface design concepts completed
5	User evaluation of prototypes in mission-relevant simulations completed to inform design
6	Human–system interfaces fully matured as influenced by human performance analyses, metrics, prototyping, and high-fidelity simulations
7	Human–system interfaces fully tested and verified in operational environment with system hardware and software and representative users
8	Total human–system performance fully tested, validated, and approved in mission operations, using completed system hardware and software and representative users
9	System successfully used in operations across the operational envelope with systematic monitoring of human–system performance

Knights & Yeates, 2021; Lever, 2011; Ralston et al., 2015; Marshall et al., 2016; Paredes & Fleming-Munoz, 2021; Thompson, 2014).

Corke et al. (1998) went on to describe progress towards vision-based control for automation of draglines and underground hydraulic manipulator arms (for a secondary rock-breaker, for example). The same Australian Commonwealth Scientific and Industrial Research Organisation (CSIRO) researchers also described progress towards automation of underground Load-Haul-Dump (LHD) vehicles and again noted the potential safety and health benefits as motivation for the work (Roberts et al., 2000).

Contemporaneous research at the Pittsburgh Research Center of the US Bureau of Mines, (Subsequently the National Institute for Occupational Safety and Health). focused on navigation for non-line-of-sight remote control of underground coal continuous mining machines with the aim of removing miners from hazardous situations (Schiffbauer, 1997). Efforts directed towards developing autonomous continuous mining equipment have continued; however, success remains elusive (Reid et al., 2011; Ralston et al., 2006, 2014; Dunn et al., 2015). That said, automated functions for continuous mining machines including heading control, and a "follow me" function that allows a flexible conveyer train being loaded by a continuous mining machine to autonomously match its speed have recently been introduced (Cressman, 2023). It has also been suggested that shuttle car automation in underground coal mines is a possibility (Corke et al., 2008). However, despite recent research on the topic (Androulakis et al., 2019; Ralston et al., 2017), this promise also remains as yet unfulfilled, as does the automated bolting required for automated roadway development (Leeming, 2023; Meers et al., 2013; van Duin et al., 2013). LaTourette and

Regan (2022) provide a discussion of the barriers facing the introduction of new technology in general in the USA underground coal mining industry, only some of which are technical.

Greater success has been achieved in the automation of underground coal longwalls using a combination of inertial navigation, lidar, and camera systems (Boloz & Bialy, 2020; Dunn et al., 2023; Peng et al., 2019; Ralston et al., 2104, 2015, 2017; Wang & Huang, 2017; Xie et al., 2018). Miners located at the face during longwall operation are exposed to major safety and health risks including explosion risks, noise, and dust exposure (e.g., Brodny & Tutak, 2018). Removing miners from the vicinity of the face via remote monitoring and automation has considerable safety and health benefits, and operating longwalls with remote supervision is becoming routine at several Australian mines (Dunn et al., 2023; Gleeson, 2021a, 2021b 2022). The implementation of "fully automated underground mining face cutting" has also been reported at a Chinese underground coal mine (Gleeson, 2023).

While technical challenges were the primary focus during the early development of autonomous trucks and LHD for underground metal mines, safety considerations also received attention (e.g., Dragt et al., 2005; Duff et al., 2002; Scheding et al., 1999; Swart et al., 2002). By 2006, automated trucks and semi-autonomous LHD were in use at the De Beers Finch underground diamond mine in South Africa (Burger, 2006). Safety of the autonomous system was achieved through an access control system that prevented unauthorised access to the automated production area. Removing system controllers from the manual loaders to a control room was noted to reduce occupational injuries.

Automated trucks and semi-automated LHD are now widely utilised in underground metal mines (e.g., Burgess-Limerick et al., 2017; du Venage, 2019; Moreau et al., 2021; Vega & Castro, 2020). The safety and health benefits of removing miners from these underground vehicles are clear (e.g., Paraszczak et al., 2015). Exposure to musculoskeletal hazards including whole-body vibration is eliminated, as are vehicle collision risks and head injuries associated with LHD buckets catching the rib while tramming. A 16–20% reduction in exposure to diesel particulate matter has been estimated to be associated with the introduction of automation to underground copper mines (Moreau et al., 2021), and loader productivity increased by about 24% (Dyson, 2020). Research towards automation of the loading phase of LHD operation has been undertaken (Tampier, 2021) and major manufacturers are offering autonomous bucket filling to allow fully autonomous LHD operation (e.g., Leonida, 2023).

Automated drills and trains are also utilised in underground metal mines, with safety and health benefits (e.g., Hariyadi et al., 2016; Li & Zhan, 2018; Quinteiro et al., 2001; Thompson, 2014; Valivaara, 2016). Automated charging of underground blast-holes has also been demonstrated (Taylor, 2023). Perhaps the most fully automated underground mine is Resolute Mining's Syama gold mine in Mali where autonomous drilling, loaders, and trucks are all utilised (Dyson, 2020).

At surface mines, dragline automation research (Winstanley et al., 2007) initially aimed to provide operator assist rather than operator replacement. The researchers recognised the importance of skilled operators and sought to augment their abilities by automating the repetitive aspects of the task. Later, Lever (2011) suggested that dragline automation was technically feasible and that its introduction was only a matter of time. However, it has become evident that removing dragline operators by automation is not being entertained (Marshall et al., 2016) nor is there a strong safety incentive to do so.

Considerable research has been undertaken towards the automation of excavators (e.g., Dunbabin & Corke, 2006; Stentz et al., 1999); however, it seems that the cost and risk of introducing automated diggers to mining operations was perceived to outweigh the benefits (Lever, 2011) and the current automation technology is restricted to "operator assist" functions (Dudley & McAree, 2016). Despite the promise of improved productivity (Yaghini et al., 2022) automated excavators and shovels do not appear to be on the horizon. There is not a strong impetus from a safety and health perspective, although technology that reduces the risks of excavators striking truck trays during loading would have safety benefits for manual truck operators.

The development of autonomous haul trucks for surface mines commenced in the mid-1990s (Nebot, 2007) with the first commercial deployments occurring in Chile and Australia in 2008. Significant cost savings, productivity improvements, and risk reductions were anticipated (Bellamy & Pravica, 2011; Lever, 2011) and these claims appear to have been achieved (e.g., GMG, 2021; Perry, 2022; Price et al., 2019). The use of autonomous haulage at surface mines has proliferated internationally across a range of commodities including iron ore, coal, gold, and oil sands. Simulation studies suggest that even greater productivity gains may be obtained by automating smaller trucks (Redwood, 2023). Water-carts are also being automated for use on surface mine sites (Westrac, 2023) as are light vehicles (Cholteeva, 2021).

Automated blast hole drilling was under development in 2006 (Lever, 2011). The first fully automated production bench drilling was achieved at a surface mine in Australia in 2014 and has become increasingly common (Morton, 2017; Onifade et al., 2023). By 2021, Rio Tinto operated 26 autonomous drills (GMG, 2021). Work is underway to automate the scanning of blast-holes and the explosive trucks that subsequently charge the blast-holes (Knights & Yeates, 2021).

Dozer automation research commenced as early as 2006 (Lever, 2011). Considerable subsequent work has been undertaken to develop a system that is capable of autonomously undertaking bulk push overburden removal (see Dudley et al., 2013; Marshall et al., 2016; McAree et al., 2017). This technology is used within Caterpillar's Semi-Autonomous Tractor System in operation at a small number of sites. Both productivity and safety benefits are claimed (e.g., Gleeson, 2021b, Theiss, 2021; Westrac, 2023). The productivity benefits are derived from increased utilisation. Removing dozer operators from the musculoskeletal hazards associated with manual dozer operation (Lynas & Burgess-Limerick, 2019) is certainly beneficial from an operator health perspective and has potential to facilitate increased workforce diversity. Removing operators from dozers undertaking high-risk tasks such as on stockpiles would also be beneficial, although in the short term this is more likely to be achieved via remote control rather than automation (e.g., Moore, 2023a; Chan, 2022).

While not strictly mining equipment, automated train loading and unloading has been in place at surface mines for some years and, after many years of effort, Rio

Tinto's autonomous train delivered its first iron ore 280 km from mine to port in 2018 (Rio Tinto, 2019). BHP commenced testing of autonomous ship-loaders at Port Hedland in 2022 (BHP, 2022a).

1.3 MINING AUTOMATION – CURRENT STATE

Based on publicly reported information, there were 183 installations of autonomous (and semi-autonomous) mining equipment fleets up to 2022 (Figure 1.1)

Australian mines hosted 44% of the installations, with Canadian mines being the next most common venue (16%). The most common fleet types were autonomous surface haul trucks and semi-autonomous underground Load-Haul-Dump vehicles, followed by autonomous surface drill rigs (Figure 1.2). The majority of Australian installations were at surface mines (64%) while the majority of Canadian installations were at underground mines (62%).

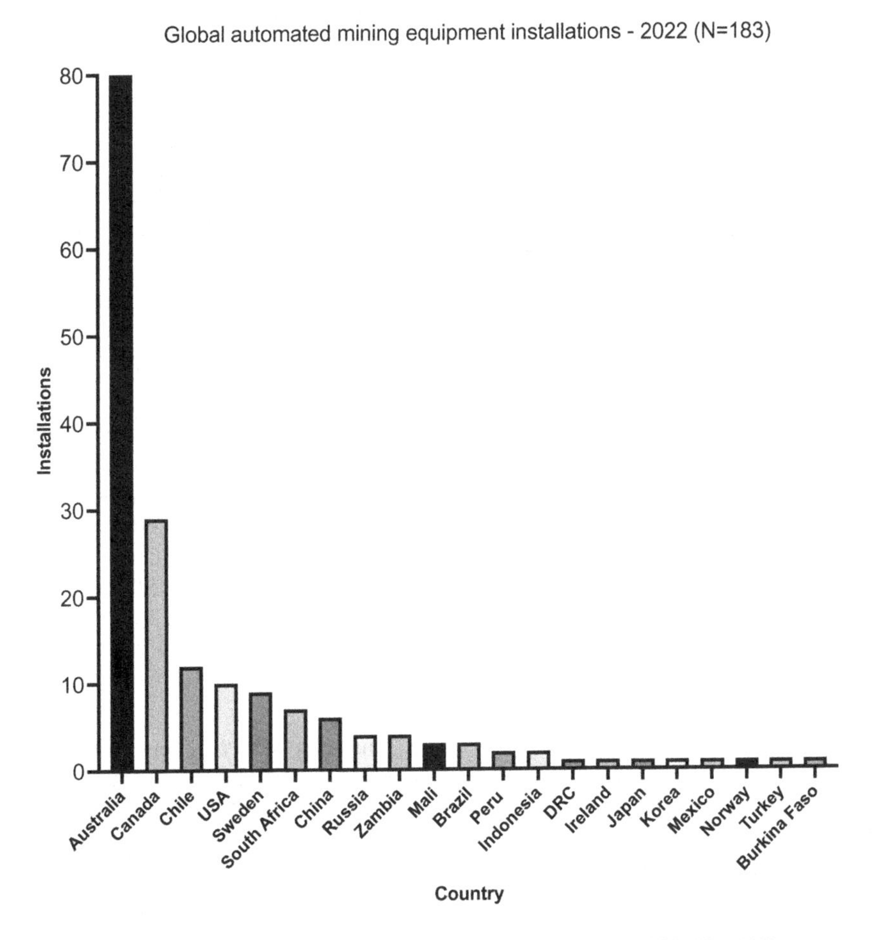

FIGURE 1.1 Global automated mining equipment installations – 2022 (N = 183)

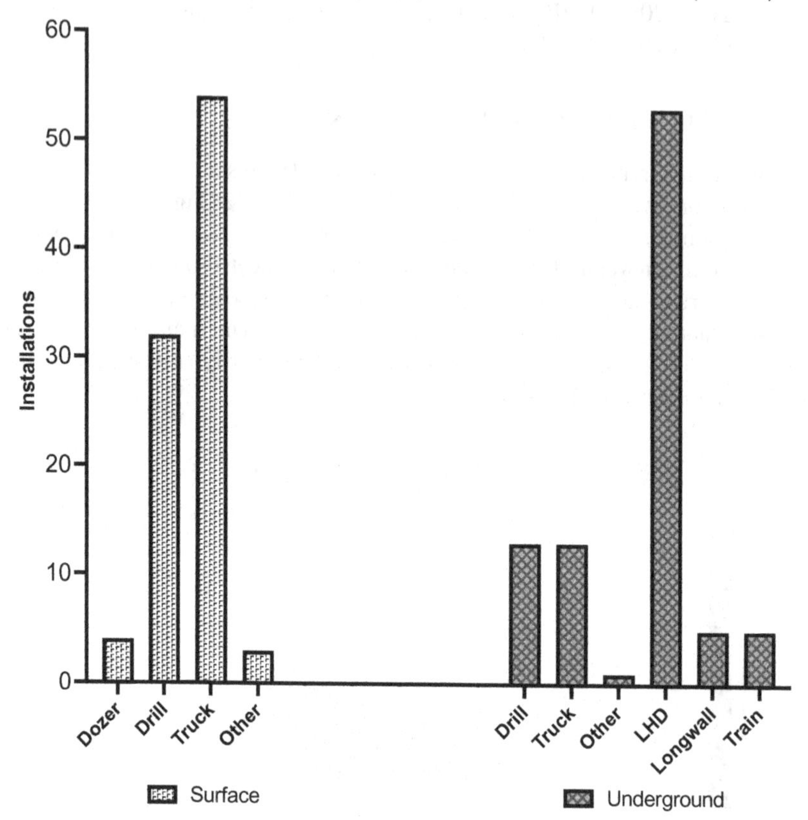

FIGURE 1.2 Types of global automated mining equipment installations – 2022 (N = 183)

The size of surface truck fleets is typically larger than for other equipment types. The total number of autonomous haul trucks in operation globally in 2022 was 1070 (an annual increase of 39%), of which 706 were operated in Australia; and the number of autonomous trucks in operation globally is forecast to exceed 1800 by the end of 2025 (FutureBridge, 2022).

1.4 MINING AUTOMATION RISKS

While automation has considerable potential to increase safety by removing people from exposure to hazards, the potential for new hazards to be introduced has also been identified (e.g., Atkinson, 1996; Benlaajili et al., 2021; Chirgwin, 2021a;2021b Lynas & Horberry, 2011; Gaber et al., 2021; Ghodrati et al., 2015; Ishimoto & Hamada, 2020; Ninness, 2018; Pascoe, 2020; Pascoe et al., 2022a, 2022b, 2022c, 2022d; Tariq et al., 2023). These hazards include new failure modes associated with sensor failure, calibration errors, software errors, communication breakdowns, or interaction between automated systems and mechanical or electrical failures. Software errors can also be introduced during upgrades, and cybersecurity risks are created. Human error associated with loss of situation awareness, mode errors, or input errors are also possible. Behavioural changes in response to the introduction of autonomous components brought about by over-trust or under-trust can compromise anticipated safety benefits. The opportunity for human supervisors within the system to be overloaded has also been noted, with potential impacts on control-room operator health. The importance of adequate training has been highlighted. Replacing field-based operators with supervisors located in a control room also means a loss of access to the information previously available to the field operators, which may contribute to delays in identifying abnormal events. The importance of well-designed and maintained haul roads has been identified as a key safety requirement for autonomous haul trucks (Benlaajili et al., 2021; Thompson, 2011).

The complexity of systems that combine automated and human elements results in situations in which adverse outcomes can arise in the absence of the failure of any element of the system Leveson (2012). The consequence is that conventional failure-based risk analysis methods may not be sufficient to understand the risks associated with the introduction of autonomous components. Hassall et al. (2022) provide examples of the use of two traditional methods (Preliminary Hazard Analysis & Failure Mode and Effects Criticality Analysis) and two alternate methods (Strategies Analysis for Enhancing Resilience & System-Theoretic Process Analysis) for identifying human-system interaction risks associated with automation in mining. Each technique identified potentially hazardous human-system interactions, and each had strengths and weaknesses. A hybrid or combination approach was suggested. Cummings (2023a, 2023b) has also pointed out that systems that utilise non-deterministic artificial intelligence or machine learning can fail in unexpected ways, and require new systems engineering processes to ensure the implications for safety are understood.

Marshall et al. (2016) identified challenges for the deployment of robotic systems in mining as including reliability, fail-safe operation with graceful failure, the design of human-machine interfaces, and safely managing the colocation of robots and humans, while Burgess-Limerick (2020) highlighted the following safety-related human factors issues associated with the introduction of automation to mining:

- Inappropriate reliance on a human “safety driver” during development or testing
- Degradation of manual skills
- Loss of situation awareness leading to delayed or inappropriate response to abnormal situations
- Nuisance alarms leading to failure to respond to abnormal situations
- Errors during human input to automated components, including mode error
- Increased span of control
- Fewer operators leading to decreased probability of abnormal event detection
- Supervisor cognitive overload

- Over-trust
- Under-trust, or deliberate circumvention of automation

These potential issues highlight the importance of ensuring that human characteristics and limitations are considered during the implementation of automation (Horberry, 2012; Horberry et al., 2018). The promised safety and health benefits of automation will only be realised if the joint system that emerges from the combination of human and automated components functions effectively.

1.5 CHAPTER CONCLUSIONS

An overview of the current state of autonomous installations worldwide was provided. The importance of ensuring that human characteristics and limitations are considered systematically during the implementation of automation was highlighted. The topic was approached from the perspective of Human Systems Integration and a scale of Human Readiness Levels widely used in other domains as a metric to track the readiness of a system for human use was presented. While automation has considerable potential to increase safety by removing people from exposure to hazards, the potential for introducing new hazards was discussed. Additionally, the complexity of systems that combine automated and human elements can result in situations in which adverse outcomes can occur where conventional failure-based risk analysis methods may not be sufficient to understand the associated risks. The next chapter provides a more detailed discussion of mining automation benefits and failure modes.

2 Automation Benefits and Failure Modes

2.1 INTRODUCTION

This chapter examines existing incident reports and other evidence including field observations to provide a summary of the benefits of automation, and the potential failure modes. We adopt the Control Framework approach developed by the Earth Moving Equipment Safety Round Table (https://emesrt.org/control-framework/). Within this framework, credible failure modes are modes of failure validated by incident experience that compromise the required operating states of the system. Evidence associated with common automated mining equipment types is collated in the following sections.

2.2 SURFACE HAUL TRUCKS

Autonomous haul trucks have been in use at surface mines for more than 10 years, providing significant cost savings and productivity benefits (Price et al., 2019). Safety has been an overriding concern of both equipment manufacturers and mining companies, and the overall collision risk profile is markedly lower than for manual truck operations. For example, an analysis of incidents associated with haul trucks, both manually operated and automated, recorded by BHP's Jimblebar mine in Western Australia for the four years that spanned the introduction of autonomous haulage to the site indicated that the overall incident rate declined by more than 90% over the period (Pascoe et al., 2022b). More recent information (Craig, 2022) indicates that the safety improvements at the site continued in subsequent years. Rio Tinto similarly has reported an order of magnitude difference in collision near-misses between autonomous and manual truck sites (Fouche, 2023).

However, an analysis of "summaries for industry awareness" provided by the Western Australian Department of Mines, Industry Regulation and Safety reveals general potential failure modes associated with automated mining equipment. Fifty-three summaries of incidents involving autonomous haul trucks reported between January 2010 and May 2021 were available for analysis.

Some of the incidents were unrelated to the autonomous functions of the truck. For example, an autonomous truck was struck by lightning:

> An empty autonomous mining truck (AMT) was ascending a ramp at an open pit when it was struck by lightning. A nearby worker witnessed a tyre exploding and causing damage to the upper structure (including the deck, autonomy cabinet, engine and cab)

DOI: 10.1201/9781003380887-2

of the AMT. … There were no injuries. Investigations found that the lightning strike initiated a chemical explosion that caused the uncontrolled deflation of the tyre.

(Incident summary SA-067-26713, 06/01/2018)

Although reported as a "potentially serious occurrence", the incident would perhaps be better characterised as a "potential serious incident avoided by automation", in that the consequences may well have been more serious if the lightning strike had occurred to a manual truck.

Mode Error. In another case, an incident occurred because a "check driver" inadvertently switched an autonomous truck into manual mode:

An autonomous mining truck travelling on the haul road in manual mode with a check driver in the cab, mounted a windrow. There were no injuries and the autonomous fleet were suspended. It appears that the check driver who was calibrating the truck inadvertently switched it into manual mode 15 seconds before the truck mounted the windrow.

(Incident summary SA-MG-453-16969, 19/06/2014)

This is an example of the general category of a "mode error" failure that can occur with any system that may be operated in different modes.

Lack of System Awareness of Environment. A relatively common incident type represented in the incident summaries is loss of traction associated with wet roads. Ten incidents were described in the database, including:

While approaching the work area of an excavator, an autonomous truck lost traction and braked causing it to slide. The road had been recently watered by a water truck. After losing traction, the autonomous truck breached the lane, attempted to correct its path and maintained its position inside the lane for ~ 45 m. The body boundary then breached the lane again when a stop event was activated on the truck. Upon braking heavily, the truck slid ~ 20 m coming to rest ~ 4 m outside of its planned lane.

(Incident summary SA-299-22131, 04/02/2016)

An autonomous haulage system (AHS) truck was travelling unloaded down a 7 degree curved ramp in an open pit, at 47 km/h, when the rear wheels lost traction against the unsealed road surface. This caused the truck to initiate medium-braking. The truck slowed to 9 km/h, while remaining in its lane, before breaching its programmed path and causing a critical braking response. The truck then slid to the left-hand side and came to rest against a windrow. The total time travelled from the initial loss of traction to rest was 9 seconds and 4 seconds passed from critical braking to rest. An initial investigation indicates the ramp was overwatered. Engineering analysis of the data recovered from the truck showed that the truck operated as designed.

(Incident summary SA-861-25701, 25/07/2017)

An autonomous surface haul truck was travelling down the mine waste ramp at an open pit when it slid and rotated about 90 degrees before rolling onto the cab side. The incident was caused by the truck moving from wet conditions on the ramp to dry as it slid.

(Incident summary SA-356-27825, 18/05/2018)

In each of these examples, although control of the autonomous truck was lost, and the truck deviated from its intended path (and in one case rolled onto its side), no other vehicles were in the vicinity. It is also notable that while the trucks may have "operated as designed", the initiation of emergency braking while sliding may not have been the optimal response to the situation.

In two further examples, the loss of control resulted in a collision with another autonomous truck:

> An empty autonomous haul truck (AHT) collided with a loaded AHT at an open pit. The empty AHT breached its lane and entered the path of the loaded AHT. Autonomous operations were suspended and an investigation commenced. It was raining heavily prior to the collision and the empty truck experienced a loss of traction.
>
> **(Incident summary SA-205-30271, 16/03/2019)**

> Following a rain event at an open pit, an autonomous haul truck made contact with the rear of another autonomous haul truck while on a pit ramp.
>
> **(Incident summary SA-275-32600, 16/02/2020)**

In both cases, the vehicles involved were both autonomous and there was no risk of injury to persons. However, it is possible that more serious consequences would arise were an autonomous truck to lose traction whilst in the vicinity of a manual vehicle.

Communications Failure. Another truck–to-truck collision occurred in February 2019. The incident summary reads:

> An autonomous haul truck (AHT) at an open pit reversed and made contact with a parked AHT.
>
> **(Incident summary SA-389-29984, 11/2/2019)**

Additional detail was provided in a news media report:

> The reversing truck stopped when communications were severed. When the wi-fi coverage returned, the truck's LiDAR (light detection and ranging) technology kicked in, detecting the presence of the truck behind it and remained stationary. … However, the truck then reversed into the stationary machine.
>
> **(https://thewest.com.au/business/mining/fortescue-metals-group-auto-haul-truck-crash-christmas-creek-no-failure-of-system-ng-b881104957z)**

Although the Chief Executive Officer of the company is quoted as saying that the incident "was not the result of any failure of the autonomous system", it appears that there was a failure of some kind involving a Wi-Fi communications error between the truck and control room (Bhattacharya, 2020). The consequences could have been serious if an occupied light vehicle had been located behind the truck at the time.

Loss of Situation Awareness. The most common type of incident described in the summaries involved interactions between an autonomous truck and another vehicle (e.g., dozer, water cart, grader, service vehicle, or light vehicle) in which the manually operated vehicle encroached into the permission line of the autonomous truck, causing the autonomous vehicle to brake. Eighteen such incidents were identified in the database, including seven in which the manually operated vehicle then collided with the autonomous truck.

One such incident was summarised as:

> An automated haul truck (AHT) turned … into the path of a manually operated water cart. The AHT was commencing a loop to position itself beneath the excavator bucket. On realising the intended path of the AHT the water cart operator commenced evasive action. However, the two vehicles collided.
>
> **(Incident summary SA-605-17670, 16/8/2014)**

Further details of the incident were subsequently provided by the regulator, as follows (Figure 2.1):

> Autonomous trucks were hauling mine waste on night shift at an open pit mine. The control room operators directed an autonomous haul truck to turn right at an intersection and perform a loop so it could be positioned under an excavator bucket on the pit floor. The intersection and turnaround loop existed in the control system but the intersection was not physically signposted on marked on the ground to alert manually controlled vehicles.
>
> A manned water cart was travelling in the opposite direction when the autonomous truck was about to turn right. The water cart driver was not aware of the autonomous truck's assigned path and, on recognising it, tried to take evasive action. The two vehicles collided, resulting in significant damage to the autonomous truck. The water cart driver received minor injuries.
>
> **(https://www.dmp.wa.gov.au/Documents/Safety/MS_SIR_226_Collision_between_an_autonomous_haul_truck_and_manned_water_cart.pdf)**

The direct causes of the incident identified by the regulator were:

- "The travel paths of the autonomous truck and water cart intersected

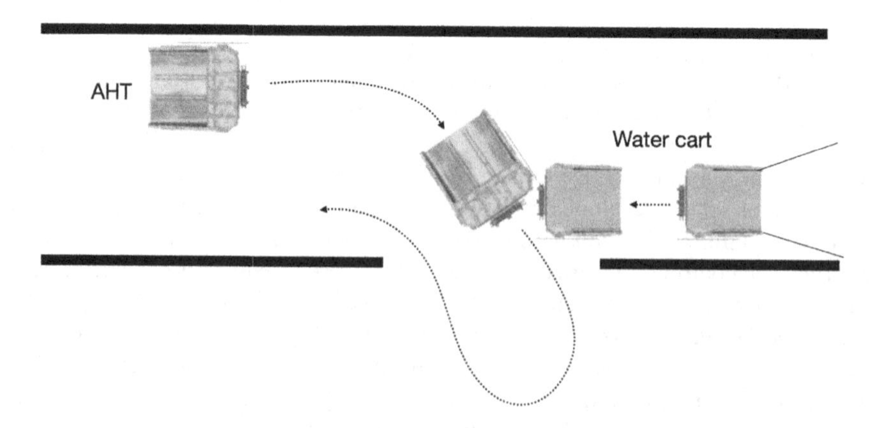

FIGURE 2.1 AHT and water cart interaction

- The turnaround loop for the autonomous truck was released for use in the control system but the corresponding intersection was not delineated on the ground, nor its intended use communicated
- On detecting the water cart in its assigned path of travel, the autonomous truck's speed (about 40 km/hr) and response time meant it could not prevent the collision."

Contributory causes identified were:

- "The change management processes for planning and assigning roads in the control system were inadequate
- *An awareness system was set up in the water cart to allow the driver to monitor the autonomous truck's path. However, at the time of the collision, the water cart driver was not fully aware of the intended path of the autonomous truck*" (emphasis added).

This last "contributory cause" identified hints at the failure mode – a loss of situation awareness by the water cart operator. The note also highlights the importance of the site awareness system provided in manually operated vehicles operated within autonomous zones.

Several other incident summaries also note the role of this interface. For example:

> A collision happened between an autonomous truck and a water cart on a ramp in an open pit. The water cart operator drove onto an active haul road while wetting a section of the pit and *observed an autonomous truck on the screen.* The operator of the water cart determined that there was sufficient room to articulate with the truck approaching and continued in the direction of travel. As the water cart came into the vision of the autonomous truck, the truck applied the brakes and began to slow down. The truck wheels then locked up and contact was made between the two pieces of equipment. The vision of the autonomous truck was impaired as the truck was approaching from the offside of the water cart.
>
> **(emphasis added; Incident summary SA-992-22337, 05/03/2016)**

> A water cart entered an intersection in the path of an autonomous haul truck during night shift at an open pit. The operator braked and came to a stop three metres from the truck. A light tower was facing the windscreen of the water cart impeding the operator's view of the intersection. *The operator used the mapping display to check on the location of autonomous vehicles and misinterpreted the location of the truck.*
>
> **(emphasis added; Incident summary SA-019-30692, 03/06/2019)**

> As the driver of a light vehicle (LV) approached an intersection on a haul road he observed the flashing light and clearance light of an autonomous haul truck (AHT). *The driver of the LV looked at the screen to view the permission line of the AHT but was unable to view it and decided to zoom out on the screen.* At that point the LV driver saw the headlights of the AHT turn towards him as the two vehicles entered the intersection. The driver of the LV applied the brakes and stopped and the AHT's

> safety systems were activated to "exception" mode (where all brakes are applied) and the vehicle stopped. The two vehicles came to rest 5–10 m apart.
>
> **(emphasis added; Incident summary SA-380-18526, 15/01/2015)**

> At a Y-intersection in an open pit a light vehicle (LV) avoidance boundary intersected the lane of an empty autonomous dump truck. The crossed path initiated a critical stop resulting in a near miss, with the vehicles coming to rest ~ 4.0 m apart. … *An investigation into the incident found that the LV driver lost situational awareness, having been distracted by focusing on the site awareness screen* located between the front seats of the vehicle out of the field of view of the driver.
>
> **(emphasis added; Incident summary SA-039-18170, 08/12/2014)**

These incidents highlight the importance of the site awareness system, and in particular the design of interfaces (Figure 2.2) provided to assist operators of manually operated equipment within the autonomous zone to maintain situation awareness. In turn, this highlights the general importance of the design of interfaces intended to provide time-sensitive information to human operators.

Over-trust. One of the "near-miss" collision incidents reported hints at a failure mode other than loss of situation awareness.

> Replays … showed a potentially serious occurrence at an open pit mine. The AHT was approaching an intersection on a haul road near the ROM, and had its permission line out, indicating its intention to turn right. As it slowed down and started turning, a light vehicle approached from the opposite direction and continued entering the intersection. The AHT identified the collision risk, applied its brakes and came to a stop. The light vehicle did not stop, but continued through the intersection, passing less than 10

FIGURE 2.2 Design of interfaces provided to assist operators of manually operated equipment within the autonomous zone to maintain situation awareness

> m from the AHT. The driver of the light vehicle failed to give way, as per pit permit requirements, and did not stop, call mayday or report the incident to their supervisor.
>
> **(Incident summary SA-520-26849, 09/01/2018)**

It is hard to imagine the operator of a light vehicle failing to notice passing through an intersection less than 10 m away from a haul truck. While this incident may have been a particularly egregious example of loss of situation awareness, it is more likely that this is an example of the general potential for "over trust" in automation to lead to behavioural changes that degrade the safety of the system – that is, the light vehicle operator had such trust that the autonomous truck would stop that they deliberately drove through the intersection in front of the truck. Combining this situation with a loss-of-traction event yields a plausible fatality scenario.

Complex Interactions. Two final summaries of automation incidents deserve comment as examples of how unwanted outcomes can arise in complex systems in the absence of the failure of any system component. The first resulted in a truck-to-truck collision:

> An autonomous haul truck stopped on an open pit ramp. A single lane was created for other autonomous trucks to pass the truck. A worker arrived to manually recover the truck. It was started and driven up the ramp into the path of a second truck as it was passing. The trucks made contact, stopping on the ramp. … The proximity detection/site awareness system was not fully operational on the first truck when it travelled into the single passing lane.
>
> **(Incident summary SA-051-28892, 03/10/2018)**

In this case, when the operator re-started the autonomous truck to drive it manually, there was a delay before the truck's site awareness system was actively broadcasting its position. No feedback was provided to the driver that this was the case and the driver had no visibility of the autonomous truck approaching from behind. This combination of circumstances resulted in the autonomous truck being unable to stop when the manually operated truck was driven into its path. All system components functioned as intended; however, the collision still occurred.

Another incident resulted in an unusual interaction between two pedestrians and the unexpected movement of two autonomous haul trucks that had serious potential consequences:

> After two autonomous haul trucks (AHTs) at an open pit lost communication, two operators were tasked with relocating the vehicles. As the first driver entered the cab of an AHT, the vehicle moved forward while the operator applied the brake and switched to manual mode. As the second operator was about to board the other AHT, its horn sounded and the vehicle moved forwards, with the operator stepping out of the way.
>
> **(Incident summary SA-743-32237, 29/12/2019)**

The regulator subsequently provided additional information. The direct causes identified were:

Operators attempted to board the AHTs while they were not under their control.

The operators did not identify that the AHTs were in exception mode when attempting to board.

Once the light vehicles in the area were deactivated, which removed the projected safety bubble, the AHTs reverted from exception to autonomous mode allowing them to resume operations.

Contributory causes were listed as:

AHTs were in exception mode and not suspended (unsafe mode to approach). Lack of understanding or clarity regarding the actions of the AHTs in various modes of operation.

Limited redundancy in communications network utilised by the AHS. Ability for personnel to override system functions that are designed as critical safety controls.

Operators did not observe the AHTs' status mode indicator lights.

Previous AHS communication issues may have desensitised the operators to potential hazards.

AHTs did not detect a person about to board.

Again, the loss of control of the situation occurred despite all systems functioning as designed. In both cases a lack of feedback to the people in the system about the state of the autonomous components, or a lack of understanding of the information provided, contributed to the event. These examples both illustrate why conventional failure-based risk analysis methods are insufficient to understand the risks associated with complex systems that include autonomous components. Additional analysis techniques such as Strategies Analysis for Enhancing Resilience (SAfER) and/or System-Theoretic Process Analysis (STPA) are also required (Hassall et al., 2022).

Input Errors. In addition to the automated haul truck-related incidents reported to the WA regulator, several other examples of incidents have been noted including:

Automation did not eliminate trucks from tipping on red lights. Mine Control were still required to remotely tip failed truck assignments. Therefore, controllers needed to observe the lighting system before overriding the truck.

(Pascoe et al., 2022b)

Although automation successfully prevented trucks from entering closed (Active Mining Areas), the system relied heavily on LV's to virtually lock the area. Driverless trucks drove into (Active Mining Areas) where light vehicles forgot to lock or engage the button effectively.

(Pascoe et al., 2022b)

These incidents are both examples of errors during input to the system.

Software Shortcomings. Another issue of concern is software change management. Considerable effort is required on behalf of mines to test the functioning of updates before installing updates because of the potential for software errors to be introduced. The extent of the effort is, in part at least, because of limited information provided by manufacturers to the mine sites about the software changes.

Control Room Situation Awareness. The autonomous trucks and associated technology, and the people in both the control room and the field, form a joint cognitive system. Timely and appropriate decision-making requires the joint cognitive system to maintain an accurate understanding of the state of the system and the environment to allow prediction of likely future events. No one person in the system has access to all the information required to maintain this situation awareness. Rather, the situation awareness is distributed across the system. Maintaining accurate distributed situation awareness is a dynamic and collaborative process requiring moment-to-moment interaction between team members and technology.

For example, the control room operator does not have direct access to information about roadway conditions and relies on people in the mine to provide the information required to allow appropriate decisions to be taken, such as slowing trucks to avoid loss-of-traction events. Similarly, the controller has access to system-wide information that needs to be communicated to field roles. Communication between team members is clearly a critical aspect of maintaining accurate situation awareness, as is acquiring and interpreting information from autonomous system interfaces. Initially, automated haulage control rooms were typically located at mine sites; however, increasingly the controllers are being moved to remote operations centres, which exacerbates this issue. The design of control room interfaces (e.g., Figure 2.3) is crucial in allowing the control room operators to play their part in maintaining situation awareness.

Musculoskeletal Injury Risks. Some limitations exist in the design of the physical aspects of controller workstations. For example, high monitor positions may lead to excessive head and neck extension and increased visual demands. Input interface requirements may also necessitate excessive use of pointing devices such as computer mice.

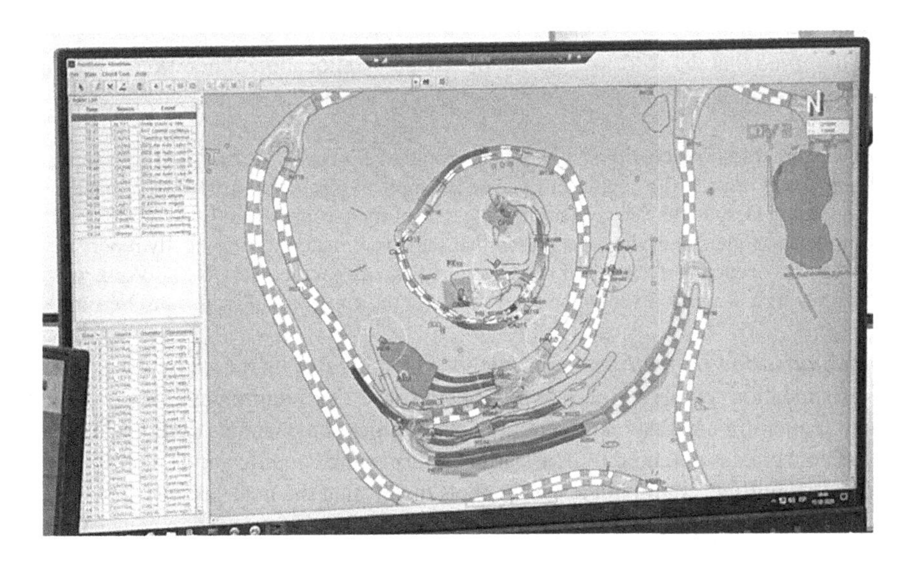

FIGURE 2.3 Design of control room interfaces

Workload. Control room roles involve high levels of cognitive workload, which may lead to performance decrements and/or adverse health effects. As well as the potential impacts on operator well-being and sub-optimal performance, there are implications for turn-over and subsequent recruitment and training costs.

For example, a controller interviewed by Pascoe explained that:

> Previously for a manned operation you wouldn't, you have 40 trucks drivers that can think about it and do it yourself. You've got one controller, on average, looking after 25 trucks, with one builder. Planning all the work for those 25 trucks, as well. So, it's constant just churn; it doesn't stop; it's relentless.
>
> **(Pascoe, 2020)**

The workload is unpredictable, and this also increases stress. As Pascoe et al. (2022b) noted:

> Supervisors can be completing monitoring tasks and simultaneously be confronted with network outages, truck slides and broken-down machines.

Interruptions to work were also noted by operators as a source of stress, for example, routine site access requests interrupting building work that requires sustained concentration.

Chirgwin (2021a) similarly noted high autonomous controller workload across multiple control rooms:

> Several controllers that had experience in manned and autonomous operations had assumed that automation would make their work life easier, but the experience was the opposite and their workload, cognitive load and communication responsibilities had increased because of automation.

and observed that the workload is also increased by allocation of additional, perhaps unnecessary, tasks:

> many organisations continued to hold on to outdated ways of working, and … continued to add tasks to the controller role. An example of this is the insistence of manual reporting. Despite the fleet systems having the ability to capture multitudes of data, all of the controllers interviewed reported that they were required to manually report on what was occurring during their shift and justifications for their actions. This task was largely seen as a task given to the controller with the aim of saving someone else time.

Communication Difficulties. Controller workload is also increased by the extent of communication required with people in field roles. The requirement for the control room to monitor and respond to multiple communication channels (radio, telephone, in-person) creates potential for frustration, interpersonal conflict, and cognitive overload. The multiple communication channels means that the field staff do not know if the control room operator is already attending to another information source. They may also not appreciate the time required to action a request before the next request can be attended to. Interpersonal group dynamics are important in this situation, particularly rapport between control room operators and field staff where interactions are largely virtual, and particularly if the controller has limited previous field experience.

The rapid expansion of autonomous haulage has resulted in mining companies encountering considerable difficulties attracting, training, and retaining controllers. This has become a vicious cycle, in that the scarcity of controllers results in high workloads, leading to burn out which exacerbates the issue. Chirgwin (2021a) described the situation she observed in multiple control rooms:

> controllers were often observed being on-shift before mining production employees, and were often the last to leave, going beyond their allocated 12hr shift. It was not uncommon to see a controller not take a break (including a toilet break) for up to 6h, and sometimes that extended to the entire shift. … Often there was no-one to replace the controller for their break, so they would either not have one, or the other controllers or their supervisor would take on the additional workload for that break period.

The shortage of controllers leads to difficulty in releasing staff for training which, in turn, also contributes to increased stress and reduced job satisfaction.

2.3 SURFACE BLAST-HOLE DRILLS

Removing operators from drill rigs removes exposure to dust and vibration, access and egress risks, and safety risks associated with vehicle travel within the mine. The advantages were described by one operation as follows:

> From our point of view in operations, what we are looking for is the precision of the process, which in drilling still depends a lot on the human factor. However, before this depended on an operator in the cabin who is exposed to risk – they are often close to the highwall, or close to bench edges or ore faces. Therefore, to remove the operator from the cabin and put them in the IROC actually improves the utilisation of the fleet while also improving the quality of life of the operator – no exposure to noise, vibration or climate extremes like cold. But it is also more efficient – for example at site the operator has a one hour lunch break, but in addition to that time they come out of the cabin, travel for maybe 30 minutes to the canteen and then the same back again. So there is unavoidable underutilisation of the drill asset. Here, the autonomous drill operator still has a lunch break but eliminates all that site related extra time … Plus the machine continues to drill anyway during lunch breaks and shift changes.
>
> **(Moore, 2023b)**

A 37% increase in drilling rate and improved accuracy, as well as increased availability, was reported for a different mine (Ellis, 2023).

Several incidents associated with autonomous drill rigs were reported in the WA incident summaries. In two cases, the rig collided with a windrow or trough; in three cases a collision, or near collision, occurred with another drill rig; and in one case the autonomous drill rig collided with a light vehicle. Where contact between vehicles occurred, the cause of the failure of the drill rig's obstacle avoidance system was not explained. For example:

> At an open pit, an autonomous mobile drilling rig was proceeding to a new drill pattern location. During the journey, the machine made contact with a parked light vehicle (LV). The drill was stopped and a supervisor informed. No injuries were sustained. The remainder of the autonomous fleet was made inactive while hazard detection systems were tested for effectiveness. An investigation was commenced.
>
> **(Incident summary SA-762-28966, 9/10/2018)**

While drill rigs are slow moving and hence the probability of a high consequence collision is low, the incident summaries highlight that obstacle avoidance technologies are fallible.

Input Errors. In another case, an input error in the location of the autonomous boundary was noted as a cause of the incident, i.e.:

> An autonomous drill boundary at an open pit was updated to allow an autonomous drill to be relocated to another area of the drill pattern. While relocating, the autonomous drill crossed the cone-delineated boundary into a manned drill area. Workers in the vicinity saw the autonomous drill behind a manned drill and called the control room operator to stop the autonomous drill tramming. It stopped about 15 metres from the manned drill. The supervisor was notified and both drills stopped work. There were no injuries and an investigation was commenced. *It was found that the updated autonomous drill boundary was incorrect.*
>
> **(emphasis added; Incident summary SA-554-27908, 27/5/2018)**

This is an example of error during input to the control system, which is a general category of potential errors associated with the introduction of autonomous components.

Different approaches to the design of autonomous drill rig workstations have been taken to allow both teleoperation and autonomous control. In some cases, the physical controls of the drill rig have been replicated in a control room.

Another approach is illustrated in Figure 2.4. Multiple video camera feeds provided a 360-degree view from the drill rig, and from a remote viewpoint, assisting the remote operator maintain global and local situation awareness. The visual interfaces previously provided within the drill cab are replicated; however, the controls located within the manual cab have been replaced by a wireless Xbox controller. The controls on the wireless Xbox controller cause different actions in each of the three modes of operation (drill mode, setup mode, and propel mode). This creates the potential for mode errors. The probability of mode errors may be reduced by ensuring that the current mode of the machine is readily apparent to remote operators. For example, auditory feedback may provide a means of identifying machine mode that does not rely on visual attention.

Operating a control in a direction which causes an effect opposite in direction to that intended is another potential error mechanism. The probability is reduced by ensuring directional control-response compatibility. Determining the appropriate directional control-response relationship is complicated in this situation because the

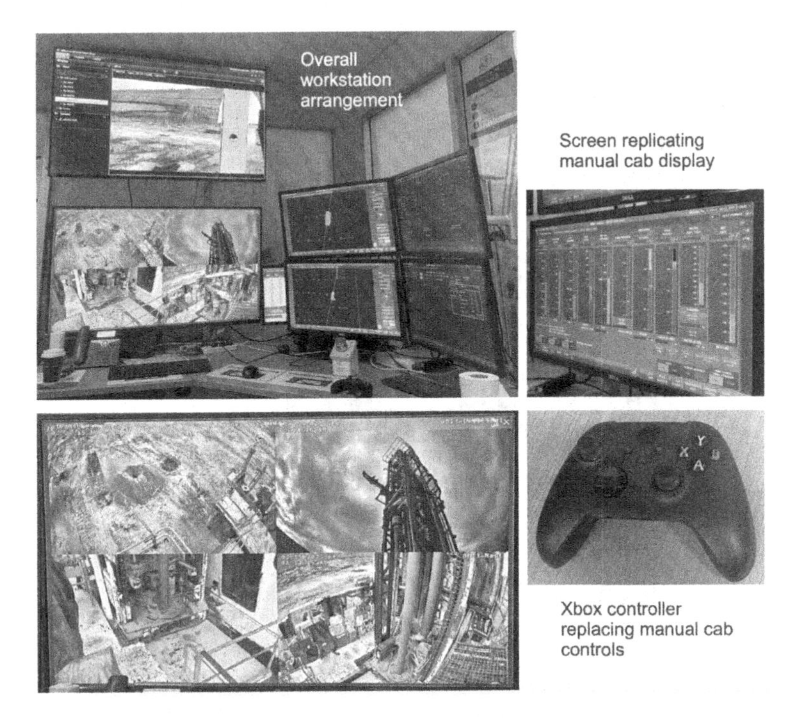

FIGURE 2.4 Multiple video camera feeds provided a 360-degree view from the drill rig

orientation of the wireless remote control may vary during use; however, there does seem to be a potential inconsistency in the directions chosen for "hoist", "jack up" in drill mode, "swing deck up" in setup mode (all upwards when the remote is in the orientation illustrated in Figure 2.5), and the control directions illustrated for "mast up" in the setup mode (the reverse).

Dozer automation has developed following the use of non-line-of-site remote control operation that was undertaken to remove human operators from hazardous areas such as stockpiles. Removing the operator from the dozer cab also eliminates exposure to whole-body vibration and musculoskeletal hazards; however, the potential for loss of situation awareness is created because of the loss of direct perceptual cues.

Video displays and a range of additional interfaces are provided to maintain situation awareness in both remote control and automated modes of operation. An extended evaluation of different combinations of visual, auditory, and motion cues for dozer teleoperation was undertaken by Dudley (2014). The intended use case was bulk dozer push at surface coal mines. Visual quality was found to be the dominant factor influencing performance while the provision of motion cues provided no additional performance benefit.

A current semi-automated dozer workstation is illustrated in Figure 2.6. Here the interfaces provided include both plan and elevation views of dozer position in addition to video feeds to aid the operator in maintaining situation awareness. In this

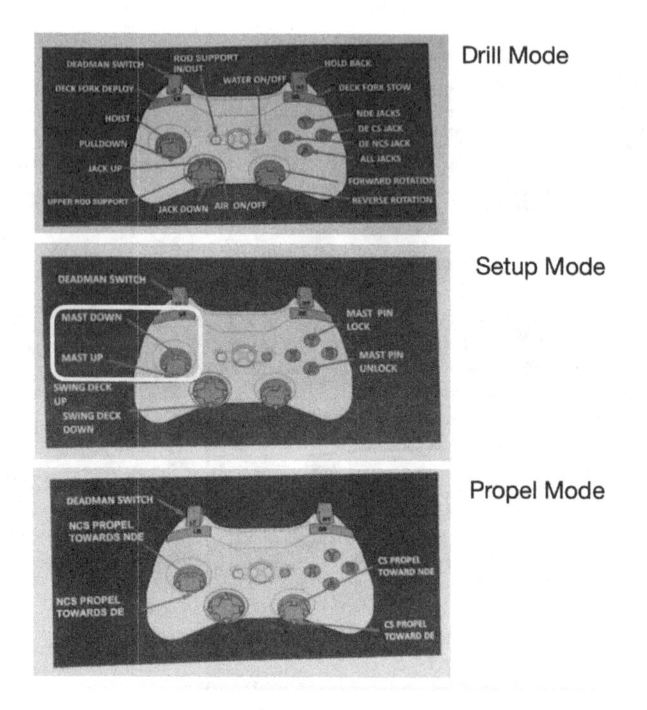

FIGURE 2.5 Determining the appropriate directional control-response relationships

FIGURE 2.6 A current semi-automated dozer workstation

case, one operator remotely supervises up to four dozers. Safety and health benefits include eliminating exposure to whole-body vibration and other musculoskeletal risk factors, access and egress, and site transport risks.

The transition to semi-autonomous dozer operation from manual required extensive operator training, starting with two dozers and gradually working up to four. Utilisation has been increased 25% and productivity is enhanced by software that automates decision-making. Alterations to the production schedule were required to take advantage of the increased equipment availability (Gleeson, 2021b).

Loss of Situation Awareness. A collision between a semi-automated dozer and an excavator occurred in 2019. The NSW Resources Regulator has provided an investigation report (NSW Resources Regulator, 2019). Semi-autonomous (SATS) dozers were being utilised to undertake bulk push operations. This technique requires an excavator to clean the rear bench material where the dozers reverse before commencing a push. The material is used to create a windrow across the back of the dozer push area. A procedural control was in place in that a manually operated machine should not operate in the active dozer slot.

Three semi-autonomous dozers were being supervised from the remote operator station by a trainee operator under instruction. Each dozer is fitted with four video cameras and these video feeds are displayed at the operator workstation. The workstation includes teleoperation controls. In semi-autonomous mode, the operator allocates a dozer to a slot and conducts the first push of the mission via teleoperation mode. The dozer then continues to operate in the same slot autonomously until either the mission is completed or until 12 passes have been conducted and the operator must reconnect with the dozer.

According to the investigation report:

> At 1.30pm, the excavator operator resumed work within the SATS avoidance zone from the north, travelling towards the south. As the edge bund was constructed using material from the highwall face, some loose material was hanging up across the face. The operator used the excavator to scale the loose material from the face, as he travelled towards the southern section of the SATS avoidance zone.
>
> As the excavator had previously scaled and cleaned up the northern area, a windrow had been built between the rear bench and the SATS dozer push slots. This resulted in the excavator working between the highwall face and the windrow. As loose material was scaled down, it was added to the windrow. The task progressed towards the south until the excavator travelled to the end of the windrow and was positioned adjacent to the rear of slot 16.
>
> At this point, dozer DZ2003 was operating in slot 16, while dozer DZ2002 and dozer DZ2010 were working in adjacent slots in the southern section of the avoidance zone about 50 metres away. Dozer DZ2003 had been operating semi-autonomously for some time. Immediately before the collision, the SATS operator had selected and was observing dozer DZ2002 until dozer DZ2010 ceased pushing. The SATS operator switched to this machine and started fault finding.
>
> Dozer DZ2003 had completed a push and was reversing towards the rear of slot 16 to start the next push. At this time, the excavator proceeded past the windrow, into slot 16. About 1.40pm, dozer DZ2003 hit the rear of the excavator. When initial contact was made, the excavator was pushed about 1.5 metres sideways, into the base of the

> highwall. The excavator then stopped sliding and dozer DZ2003 continued to tram in reverse, colliding with the excavator multiple times trying to reach its programmed GPS coordinates.
> Dozer DZ2003 eventually lost traction and after five seconds, the control system faulted and stopped tramming. From the initial contact to dozer DZ2003 stopping was about 14 seconds. The excavator had some damage however the operator was uninjured.

When a dozer is selected by the supervisor, a screen in front of the operator displays the four cameras corresponding to the dozer that is the focus of the operator's attention. A small side panel also shows two camera views for each of the other three dozers. While information was available to the supervisor, it was not provided in a way that facilitated maintenance of accurate situation awareness. Humans are very poor at vigilance tasks (Horberry et al., 2011). It is entirely understandable that the impending collision was not identified by the supervisor who was focused on fault-finding on a different dozer.

2.4 UNDERGROUND COAL LONGWALL

Working at the longwall of an underground coal mine is associated with a range of safety hazards, most notably rock falls, outbursts, or the ignition of methane. Health hazards, and particularly exposure to respirable dust and noise, are also associated with working in the area.

Automation has great potential to reduce the exposure of miners to these hazards. Current technology has removed two miners to a surface control room. While the majority of the crew remain underground, they work in less hazardous locations. Remote guidance technology continuously steers the longwall, automatically plotting its position in three dimensions and allowing real-time monitoring of progress. Control room interfaces (Figure 2.7) provide video feeds and other information to compensate for the loss of direct perceptual information. According to the CSIRO, longwall automation technology has increased productivity by 5–10% through improved consistency.

Credible failure modes associated with longwall automation include loss of situation awareness resulting from the loss of direct perception including vibration of machine and auditory information from cutting heads; loss of manual skill; communication technology disruptions; communication difficulties; high cognitive workload; and musculoskeletal injury risks associated with sedentary work.

Rather than relocating some crew members permanently from underground to the control room, the miners rotate between the surface and underground on different shifts. This is beneficial in rotating exposure to the physically sedentary but cognitively demanding control room work across miners as well as maintaining underground knowledge and skills. While potentially decreasing safety and health risks, further automation will reduce these rotation opportunities.

Although the control room interfaces provide extensive information sources and are well designed, it was noted by operators that additional camera views would be beneficial and that communication between the surface control room and underground workers at the longwall was difficult at times.

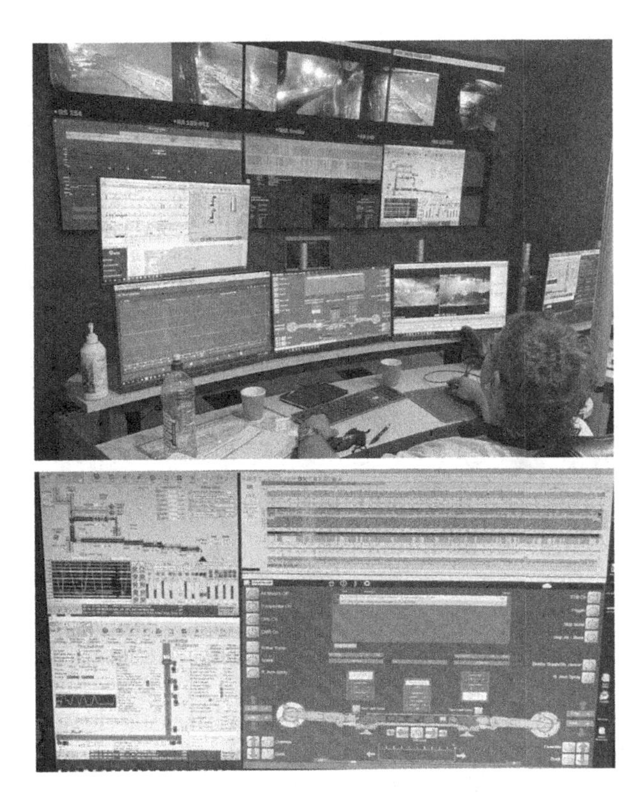

FIGURE 2.7 Control room interfaces for an automated underground coal longwall

2.5 UNDERGROUND LOAD-HAUL-DUMP VEHICLES

Semi-automated Load-Haul-Dump vehicles (LHD) have been installed at more than 50 mines globally since 2006. Operators located in a control room load the LHD bucket using tele-operated control. The loader is then switched to autonomous mode to travel to the dump where the load is dumped autonomously. The loader then autonomously returns to the next load point selected by the operator. Operators may be responsible for supervising multiple loaders. A range of interfaces are provided to allow the remote operator to maintain situation awareness and remotely control the loading phase (Figure 2.8).

The safety and health benefits of removing miners from these underground vehicles are clear. Exposure to musculoskeletal hazards including whole-body vibration is eliminated, as are vehicle collision risks and head injuries associated with LHD buckets catching the rib while tramming. A 16–20% reduction in exposure to diesel particulate matter has been estimated to be associated with the introduction of automation to underground copper mines (Moreau et al., 2021).

Unauthorised Access to Autonomous Zones. No reports of injuries associated with autonomous loaders have been identified. This is likely to be, in part at least, because the current practice is to isolate all other equipment and pedestrians from the

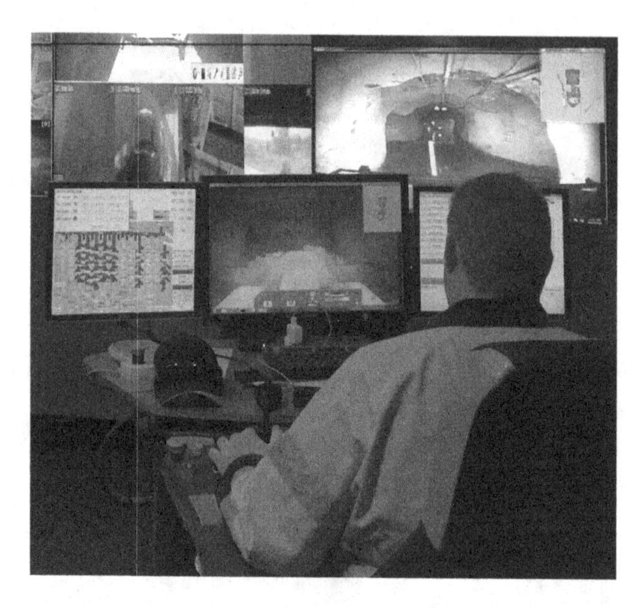

FIGURE 2.8 Interfaces to allow the remote operator to maintain situation awareness and remotely control the loading phase of a semi-automated underground metal mine LHD

zones in which autonomous LHDs operate. However, incidents have occurred where automated equipment has been activated in an isolated autonomous area with multiple working faces while persons were located at another face (Thompson, 2023).

Remote-control loaded buckets, on average, contain smaller loads than manually loaded buckets. However, overall productivity is higher because increased equipment utilisation arises as a consequence of being able to continue operation through blasting and shift changes. LHD supervisors located remotely from the loaders may not feel the quality of the roadway and allow the loaders to drive at speeds that increase unplanned maintenance requirements.

The location of the control workstation for the semi-autonomous LHDs varies across sites. Some sites locate the LHD control room underground and this was seen as beneficial for maintaining communication with other staff underground. Other sites have chosen to locate the LHD control room on the surface of the mine to reduce underground travel time, or in a remote operations centre in a city location at some distance from the mine. At one mine, semi-autonomous LHD supervision is undertaken both in a control room on the surface of the remote mine and in a control room located in the company's city office. This additional location allows staff who are unable to undertake a Fly-In-Fly-Out roster to continue to work for the mine.

The integration of semi-autonomous loaders into existing production systems is not straightforward and sites noted that difficulties typically arise in maintaining production during the transition. Not every site that implemented semi-autonomous LHDs has persisted with the technology and some sites have taken several attempts before being successful. Implementation of LHD automation requires a strong mandate from the highest levels of the company to be successful in the face of inevitable, if temporary, production declines during the implementation phase.

This observation is consistent with the findings of a case study of the successful implementation of semi-autonomous loaders at NorthParkes. The strategies for successful automation implementation included involving all people who will be impacted; encouraging constant communication between operators and designers; providing operators with essential information; avoiding to provide non-essential information; providing the operators with flexibility; empowering operators to take action; and taking advantage of the new possibilities provided by automation (Burgess-Limerick et al., 2017).

One issue identified at NorthParkes during the initial preparation for the transition to autonomous LHDs was that all underground tasks would be affected by the change. For example, at shift change the continued operation of LHDs from the surface enables production to continue, removing time pressure and allowing greater time for shift handover. However, it was also identified that access to, or through, sections of the mine where autonomous loading was in operation would be prevented and this impacted the performance of many other tasks.

Constant communication between operators and designers throughout the implementation and subsequent operation of the semi-autonomous loaders was critical in developing and refining the control-room user interface. Continuous presence of manufacturer expertise on-site allowed a rapid feedback loop with designers.

Providing operators with opportunities to suggest modifications to the system was a key feature in the success of the implementation. Operators continually updated a list of issues and a "wish list" of improvements, which were fed back to the system designers, and many changes resulted. For example, equipment damage was occurring because the loader was hitting the walls of the draw point while under manual control. Using the laser scanners already in place for autonomous navigation to detect the proximity of the walls was suggested during manual operation and to convey this information to the operators through changes in colour of the scanning information provided on the teleoperation assist window. This information was also used to automatically apply the brakes if necessary to prevent collision with the walls.

Similarly, wheel spin caused damage to the LHD wheels but was hard for operators to detect while loading remotely. A wheel-slip detection sensor was added and an indication of wheel slip was provided to the operator through a change in colour of the schematic loader wheels in the teleoperation assist window. In both cases, the presentation of relevant information to the operators in a meaningful way ensured the information could be used effectively to reduce equipment damage.

Relevant information is also conveyed inadvertently, rather than by design. One operator explained that it can be difficult to gauge when the bucket has been lowered sufficiently to the ground in preparation for loading; however, if too much pressure is placed on the ground by the bucket, the front wheels will raise and the wheels slip. The operator noted that the camera shake, which could be seen on the video feed when the bucket was lowered, was a useful cue.

Conversely, another change made during system implementation was to reduce the number of fault alarms presented to the operator. Many of these alarms, while relevant to an engineer during commissioning, were not relevant to the day-to-day operation of the LHD. As well as being a nuisance to operators because each message required acknowledgement, becoming habituated to frequent non-essential error messages was reported to have led on at least one occasion to an operator failing to react to a critical error, with potentially serious consequences.

Providing flexibility in information provision was another strategy employed. The LHDs are fitted with a microphone and the audio is available to the operators; however, it was found that this information was not wanted by the operators and the audio is left off because the nuisance value of the noise outweighed the benefit of any relevant information conveyed.

Many details of the automation implementation were left to production crews to determine. For example, in the transition to autonomous loading, some crews decided that crew members would be trained for autonomous control, while other crews chose to have specialist autonomous operators. The number of LHDs for which an operator should have responsibility was also determined by the crews. While four loaders can be controlled by one person, the cognitive load was overly fatiguing and three was determined to be the typical maximum. During operation, some crews chose to allocate three LHDs to be controlled by each operator, while other crews allowed more flexibility, with all loaders able to be controlled by any of the three operators on shift at any one time.

Allowing crews to choose different strategies provides opportunity to evaluate different options, and comparisons between operator and crew productivity can be used to fine-tune operator strategies and identify aspects of operator behaviour which lead to improved productivity. Production crews have also taken action without involving the system designers. One issue encountered was that the cameras and scanners were accumulating dust which was causing the automation to fail. While the system designers were exploring options for on-board cleaning mechanisms, the crews devised a means of dumping water on the camera and scanners when required. Making all aspects of the control system as flexible as possible and giving operators maximum control over the automation increases the opportunities that operators have to adapt to new situations.

The implementation of autonomous loading has also had unanticipated consequences for future process improvements. The ability to more flexibly execute different draw-point extraction patterns, and modify these extraction patterns, prompted the development of optimisation software to determine in real-time the optimal pattern of extraction. This is itself a form of automation which will provide assistance to the shift boss in maintaining situation awareness of the extraction and aid decision-making.

2.6 CREDIBLE FAILURE MODES

The information reviewed in this chapter has revealed that the introduction of automated components introduces new failure modes that have potential for adverse safety and health outcomes. These generic failure modes include:

- ***Software Shortcomings***. It is difficult to verify that software is trustworthy. Testing can only reveal the presence of flaws rather than prove the absence of errors. This is particularly true if machine learning is involved. Operations that have implemented autonomous machinery described spending considerable time verifying the operation of software updates prior to release.
- ***Communication Technology Disruption***. Autonomous mining systems are dependent on continuous digital communications. Considerable effort is required to ensure the required networks are in place and maintained. Loss of network connectivity is a common cause of lost productivity and at least one potentially serious incident has occurred in which a communication interruption was implicated.
- ***Cyber Security Breach***. Breaches have occurred and this is a risk that will increase. Continuous attention to network security is warranted given the potential damage that a malicious actor could achieve. The human aspects of cyber security also require attention.
- ***Unauthorised Access to Autonomous Zones***. Incidents have occurred at surface mines where vehicles not fitted with site awareness systems have accessed active autonomous zones without an escort. In the underground context, incidents have occurred in which automated equipment was activated in an isolated autonomous area with multiple faces while persons were located in the area.
- ***Loss of Manual Skill***. Machine operators' manual skills will deteriorate if not practiced. Whether this is a concern will depend on whether the system concept of operation includes reinstating manual operation at any time, and in what circumstances.
- ***Over-trust***. People working in the vicinity of autonomous systems are likely to change their behaviour to take advantage of the perceived safety features of the system. Driving a light vehicle through an intersection in front of an autonomous truck, and trusting that the truck will take evasive action, is an example. Ensuring that people working with autonomous components have an accurate understanding of the system's capabilities and limitations, and the physical constraints, is important. So is supervision, monitoring, and enforcement of safety-related procedures such as hierarchy road rules.
- ***Input Errors***. Whenever human controllers are responsible for entering information into the system there is potential for error. The probability of such errors is reduced by effective software and interface design. Where remote control is included in the concept of operations, the design of the workstation controls should take into account the possibility of mode errors and ensure that directional control-response compatibility is maintained.
- ***Inadvertent Mode Changes***. Whenever equipment can be operated in different modes there is potential for inadvertently switching between modes. This includes switching between autonomous and manual modes.
- ***Complex/Emerging Interactions***. Systems including autonomous components may give rise to unpredicted adverse consequences even when all components function as intended. The use of systems-based risk analysis

techniques such as Systems-Theoretic Process Analysis (STPA) is recommended to identify and control such potential outcomes.

- ***Sensor Limitations***. Sensors have limitations that can result in the system losing awareness of the situation. These limitations require analysis and management.
- ***Lack of System Awareness of Environment***. Removing operators from direct perceptual contact with the operating environment creates the potential for loss of awareness of the environment. One example is wet roadways leading to loss of traction.
- ***Loss of Situation Awareness***. Several incidents have occurred in which the operators of equipment being operated manually in the vicinity of autonomous haulage have failed to predict the movements of the autonomous haulage, despite being provided with a system interface intended to provide this information. Incidents have also occurred in which an unfolding situation has not been identified by a control room operator despite, for example, video feeds providing the necessary information. These incidents highlight the difference between information being available and being perceived, and hence the critical importance of interface design to assist the people within the systems to understand current system states and accurately predict the likelihood of future states.
- ***Distributed Situation Awareness Challenges***. A related issue is that in many systems, there will be no individual who possesses all the information required to maintain overall situation awareness of the whole system. Instead, the situation awareness is distributed across the people and technology within the system. Maintaining accurate distributed situation awareness is a dynamic and collaborative process requiring moment-to-moment interaction between team members and technology that can be hindered by limitations in system, or interface, design.
- ***Communication Difficulties***. Communication between team members is critical. Difficulties associated with technology limitations, or cognitive overload caused by multiple simultaneous communication channels, can impede performance with potential safety or health consequences. Non-technical skills, and the absence of psychosocial hazards, are also required to ensure effective teamwork.
- ***Workload***. Potential exists for control room operators or others impacted by the introduction of automation to be overloaded, with consequential risks of errors, and adverse health consequences. The workload of all people within the system is a key aspect for consideration in system design.
- ***Musculoskeletal Injury Risk Factors***. Long-duration sedentary work with few breaks combined with static or awkward postures and/or excessive pointing device use, especially if accompanied by psychosocial risk factors such as high cognitive workload, time pressure, and/or conflict with peers or supervisors may create a situation in which musculoskeletal injury risk is high.

Effective risk management requires analysis of these potential unwanted events during system design. The analyses undertaken should include task-based risk assessments involving a range of operators and others effected by the system, and systems-based techniques, in addition to conventional hazard-based risk analysis techniques. All of these failure modes involve human interactions with the technology. The risks should be reduced during a human-centred system design process that focuses on the role, capabilities, and limitations of the people in the system. Residual risks need to be understood by mine management to allow effective controls to be devised, implemented, and monitored.

2.7 CHAPTER CONCLUSIONS

Benefits and failure modes associated with common autonomous mining equipment (surface haul trucks, blast-hole drills and dozers, and underground longwall and load-haul-dump vehicles) were discussed. Credible failure modes were identified, including software shortcomings, communication technology disruption, cyber security breach, unauthorised access to autonomous zones, loss of manual skill, over-trust input errors, inadvertent mode changes, complex/emerging interactions, sensor limitations, lack of system awareness of the environment, loss of situation awareness, distributed situation awareness challenges, communication difficulties, workload, and muscular injury risk factors. All of the failure modes identified involve human interactions with the technology. The following chapter provides examples of best practice in automation deployment gathered from other industries.

3 Best Practice in Other Industries

3.1 INTRODUCTION

This section describes Human-Centred Design and Human Systems Integration and shows their value to the mining industry. The definitions of Horberry, Burgess-Limerick, and Steiner (2018) are used throughout. The chapter also summarises how human-centred design and human systems integration complement and relate to each other.

Human-Centred Design is widely used in the design of medical equipment, road vehicles, and consumer products. It is a process that aims to make equipment and systems more usable and acceptable by explicitly focusing on the end users, their tasks, their work environment, and the use context. A key requirement of human-centred design is for users and other stakeholders to be involved throughout the design and development of the equipment or system. However, to date, human-centred design has not been widely applied to the design, development, and deployment of equipment or new technology in the mining industry.

Human-centred design can result in fewer operator errors, decrease training costs, promote better system usability, and lead to improved operator acceptance of systems. In mining, it can help minimise design issues like operator overload from too many warnings in truck cabs, or self-rescue equipment not being deployed because of poor device usability. The most relevant ISO standard in this area (ISO 9241 Part 210: Human-centred design for interactive systems, 2019) is now becoming well accepted in many domains. It can easily be applied to mining equipment and new technologies.

Originally conceived in the context of defence procurement, human systems integration is a broad overarching framework applied originally within defence and increasingly within other industries to ensure that total system performance is optimised and total operating costs are minimised. Booher (2003) defined human systems integration as the process of integrating the domains of human factors engineering, system safety, training, personnel, manpower (crewing), health hazards, and survivability into each stage of the defence-systems-capability life cycle (needs, requirements, acquisition, service, and disposal).

Human-centred design as described above for the minerals industry refers to a process by which some of the key goals of human systems integration are achieved, with a particular focus on the domains of human factors engineering, systems safety, health hazards, and training. Considering the broader context is important as it shows that safety and performance in complex mining systems can impacted by decisions

DOI: 10.1201/9781003380887-3

and actions made at all levels of the system, not just by operators and maintainers working with equipment at the sharp end.

In this section, leading practice human-centred design and human systems integration will be presented in the following domains: defence, aviation and space, road transport, rail, and healthcare.

3.2 DEFENCE

3.2.1 Overview

Human systems integration refers to a set of systems engineering processes (INCOSE, 2015) originally developed by the US Defense Industry (Booher, 2003), which ensure that human-related issues are adequately considered during system planning, design, development, and evaluation. Defence agencies typically have formal human systems integration policies in place for major systems acquisition. While the details vary across agencies, the general procedure is to place responsibility on the programme manager to ensure that implementation of human systems integration occurs during equipment acquisition.

For example, US Department of Defense Instruction 5000.02 "Operation of the Defense Acquisition System" (2008) required the programme manager:

> to have a plan for HSI in place early in the acquisition process to optimize total system performance, minimize total ownership costs, and ensure that the system is built to accommodate the characteristics of the user population that will operate, maintain, and support the system.

Further, Instruction 5000.95 "Human Systems Integration in Defense Acquisition" (2022) assigns responsibilities and prescribes procedures for achieving this across the domains of human factors engineering; personnel; habitability; manpower; training; safety and occupational health; and force protection and survivability.

Programme managers are required to undertake a combination of risk management, engineering, analysis, and human-centred design activities including:

(i) the development of a human systems integration management plan
(ii) taking a human engineering design approach for operators and maintainers
(iii) task analyses
(iv) analysis of human error
(v) human modelling and simulation
(vi) usability and other user testing
(vii) risk management throughout the design life-cycle
(viii) developing a training strategy

In the Human Factors Engineering domain, the programme manager is required to ensure that design considerations addressed include the design and layout of work environments; human-machine interfaces; design for maintenance; automation; and minimising system characteristics that require excessive cognitive, physical, or sensory skills, or workload-intensive tasks.

The Personnel domain primarily involves identifying the knowledge, skills, and abilities of available personnel, while Manpower considers the requirements of the system. Activities in the Training domain include training effectiveness evaluations to develop options for individual, collective, and joint training activities including simulation-based training. Habitability concerns the requirements for the physical environment; the Safety and Occupational Health domain focuses on minimising health risks; and the Force Protection and Survivability domain concerns the mitigation of external treats.

The processes described ensure that human considerations are integrated into the system acquisition process. Human performance is considered to be a key factor in "total system performance" and it is recognised that enhancements to human performance will correlate directly to enhanced total system performance and reduced life cycle costs.

The UK Ministry of Defence refers to "Human Factors Integration", rather than human systems integration; however, the intent conveyed by Defence Standard 00-251: Human Factors Integration for Defence Systems (UK Defence Standardisation, 2015) is similar. The formal requirements are set out in Joint Service Publication 912 (UK Ministry of Defence, 2021) "Human Factors Integration for Defence Systems". Here the domains are identified as Personnel; Training; Human Factors Engineering; System Safety & Health Hazards; and Social & Organisational.

Part 1 of Joint Service Publication 912 requires that the following "shall be fully pursued to achieve satisfactory outcomes" in all acquisition projects:

a. "ensure that all people-related Risks, Assumptions, Issues, Dependencies and Opportunities (RAIDO) are identified and managed from the very outset of a project, and throughout the rest of life cycle.
b. ensure that all Human Factors Process Requirements (HFPRs) are specified, thereby assuring that HFI processes are properly and adequately undertaken.
c. ensure that Human Factors System Requirements (HFSRs) are specified, thereby assuring that people-related technical aspects of the Solution are properly and sufficiently addressed (based on the identified RAIDO).
d. ensure that a human-centred design approach is adopted, involving the End Users in system and equipment design and evaluation.
e. ensure that established Human Factors principles, accepted leading practice, and suitable methods, tools, techniques, and data are used.
f. ensure that the HFI programme is designed to align and integrate effectively with the project life cycle.
g. ensure that people-related considerations of the Solution undergo formal scrutiny, assessment, and acceptance."

Part 2 of Joint Service Publication 912 provides guidance in achievement of these goals. It is noted that failure to consider humans can increase the risk of accidents, increase costs (including training costs), and reduce performance. Human factors integration activities are defined to include: analysis activities (requirements analysis; task analysis; human performance modelling; human reliability analysis; training needs analysis); design activities (application of standards; modelling, prototyping, human computer interface design, workplace design, training design, organisation design); and test and evaluation activities (assessment of compliance, manual handling assessment, safety case, assessment of procedures, training evaluation).

3.2.2 Defence Examples

F-119 engine maintainability. During the process of competing for the contract to produce an advanced tactical fighter (F22 Raptor), Pratt & Whitney focused their strategy on reliability, maintainability, and sustainability (Liu et al., 2010). Of the order of 200 studies were undertaken to examine criteria across the human systems integration domains. The final requirements included: only five hand tools utilised to service the engine; all line replaceable units were able to be serviced without replacing any other unit; each unit is replaceable using a single tool within 20 minutes; and maintenance is possible while wearing hazardous environment protection clothing. Importantly, the extensive commitment by the manufacturer to improving maintainability was a direct consequence of the emphasis placed on this issue by the US Air Force during the acquisition process and was central to the manufacturer's competitive strategy.

Guided missile destroyer bridge mock-up. A full-scale mock-up of the design of a naval destroyer bridge was used with multiple teams to assess performance across different scenarios. A range of design deficiencies were identified at very early design stages that were able to be corrected while still drawings. It was suggested that a $20,000 investment in this process achieved cost avoidance of the order of $20 million (Hamburger, 2008).

3.2.3 Implications for Mining

The key learning from the defence industry for the mining industry is the importance of considering human systems integration as a fundamental component of the acquisition/procurement process. The development of a human systems integration management plan is a crucial early step in procurement.

3.3 AVIATION AND SPACE

3.3.1 Overview

Predating the development of human systems integration as a formal discipline, the USA Federal Aviation Administration (FAA) Order 9550.8 (FAA, 1993) required that "Human factors shall be systematically integrated into the planning and execution of the functions of all FAA elements and activities associated with system acquisitions and system operations". This is required at the earliest opportunity to achieve increased performance and safety, and decreased lifecycle staffing and training costs. The FAA (2016) standard HF-STD-004a titled "Human Factors Engineering Requirements" describes human factors activities expected to be undertaken by vendors and requires that the activities to be conducted are described in a Human Factors Program Plan, including tasks to be performed; human factors milestones; level of effort; methods; design concepts; and human factors input for the test and evaluation plan. Human factors collaboration with other disciplines including safety, training, systems engineering, and personnel selection is required.

The USA National Aeronautics and Space Agency (NASA) requires Human Systems Integration to be implemented and documented in a Human Systems Integration Plan. The plan identifies the steps and metrics to be used throughout a project life-cycle, and the methods to be undertaken to ensure effective implementation and maximise system performance and safety while reducing risks and life-cycle costs. A "Human Systems Integration Handbook" (NASA, 2021) provides guidance.

Human systems integration is defined by NASA as the:

> interdisciplinary integration of the human as an element of a system to ensure that the human and software/hardware components cooperate, coordinate, and communicate effectively to successfully perform a specific function or mission.

The system is defined to include hardware, software, and humans, as well as data and procedures. The human systems integration plan must be established early in programme planning and the processes undertaken iteratively, considering all people who interact with the system throughout the entire life-cycle, and collaborating across multiple domains.

While the origins of the discipline in defence are noted, the relevant domains for application to aviation and space are defined by NASA as:

- Human Factors Engineering – Designing and evaluating interfaces and operations considering human performance characteristics. Activities in this domain include analyses of tasks and human performance capabilities and limitations, and evaluation of design alternatives.
- Operations – Life-cycle engagement of operational considerations into design for human effectiveness for operations and maintenance crews. Particular attention is directed to the design of interactions between human and automated components of the system.
- Maintainability and Supportability – Designing for simplified maintenance to reduce maintenance errors, as well as increasing maintenance efficiency (reduced training and manpower) and system availability.
- Habitability and Environment – Design of living and working environments including lighting, ventilation, noise, temperature, and environmental health.
- Safety – Life-cycle consideration of safety to reduce risks. Activities in this domain include systematic analyses of risks and development of system

designs that minimise these risks. Attention is directed to both safety and health hazards.

- Training – Design and implementation of training to equip all humans in the system with the knowledge, skills, and attitudes required to accomplish mission tasks. It is noted that training planning should occur throughout the project life-cycle because design decisions will impact the extent and nature of training required. Analyses of training needs provide input to the evaluation of design alternatives.

Human systems integration is defined to include analysis, design, and evaluation of requirements, concepts, and resources across the domains. Verification requires a combination of human-centred testing, modelling, and analysis.

Effective application of human systems integration is understood to result in improved safety and health, increased user satisfaction and trust, increased ease of use, and reduced training time; all leading to higher productivity and effectiveness. Conversely, NASA identifies failure to apply human systems integration as increasing risks of major accidents as well as minor incidents, greater training requirements, and higher costs including those associated with redesigns and maintenance. The NASA handbook provides a series of case studies providing examples of the value of effective human systems integration, and examples of the consequences of failing to do so.

The European Organisation for the Safety of Air Navigation (Eurocontrol), which develops standards for Air Traffic Management (ATM) systems and services, has published a white paper titled "Human Factors Integration in ATM System Design" (2019) that provides principles for the integration of human factors and ergonomics (HF/E) in system design with a particular emphasis on the achieving the anticipated benefits of automation. These are summarised as:

1. "Build **joint design teams** and do not treat HF/E as a mandatory add-on
2. Make a coherent **user-centred-design rationale** your HF/E product
3. Strive for a short, iterative **user-centred design process**
4. Derive **objective HF/E criteria** instead of relying on user opinions
5. Evaluate as early as possible with the help of **prototypes**
6. Select appropriate **conditions for evaluation**: Evaluate day-to-day operations as well as critical situations
7. Support the **problem-solving** process during implementation by facilitating trade-offs
8. Do a proper **problem-setting** in the first place whenever possible to understand your actual problem and the underlying mechanisms and needs
9. Be ready to participate in strategic decisions and introduce a **purpose-orientated view of Technology**" p. 3 (emphasis in original).

The Eurocontrol white paper highlights the user-centred design process defined by ISO 9241-210. and suggests that this must become standard practice.

3.3.2 Examples

Baggerman et al. (2009) provide examples of successful human systems integration implementation in civilian aerospace, including references to historical successes of human factors engineering in the Apollo program, as well as the Constellation program's Crew Exploration Vehicle, Lunar Lander and extra-vehicular systems. Several notable failures to adequately undertake human systems integration are also noted in the case studies provided in the NASA handbook, including the Boeing 737Max.

3.3.3 Implications for Mining

Human systems integration principles and particularly the use of human-centred design methods should be applied to procurement of all levels of complexity and scale. Doing so is crucial to achieving the anticipated benefits of automation.

3.4 ROAD TRANSPORT

3.4.1 Overview

The globally leading approach to manage human and technical risks in the road environment is called the "Safe System". It was originally adopted in European states such as the Netherlands and Sweden, but has more recently been deployed successfully in Australia, the UK, and New Zealand (PACTS UK, 2022). In combination with wider human systems integration aspects, human-centred design has long been recognised as beneficial in road transport, particularly by car original equipment manufacturers (OEMs), and is often used as a point of differentiation to other manufacturers.

In the past 70 years there have been markedly different paradigms to road safety in Australia. In the 1950s, the focus was on the road user – particularly to blame the victim. In the 1960s and 1970s the leading approach focused on systemic interventions such as the "Haddon matrix" that examined crashes in terms of system elements such as the driver and the road environment and stages in the accident sequence. In the 1980s and 1990s the focus was on top-down aspects such as targeted national plans and highway agency accountability. From the 2000s onwards, the Safe System approach of shared responsibility and error tolerance has become the key approach.

3.4.2 The Safe System Approach

The Safe System approach holds that the responsibility for crashes and injuries is shared between the designers, providers, regulators, and users of the road transport system. The long-term vision is for the road transport system to be free of deaths and serious injuries. It holds that sustained innovation is required to proactively build safety into the road transport system, rather than reactively adjusting to system failures. A safe system in Australian road transport depends on understanding and implementing the following key guiding principles:

- Shared responsibility: this means all stakeholders take an individual and shared role in road safety. As well as actual road users, these include pedestrians, planners, engineers, parents, policymakers, enforcement officers, educators, utility providers, insurers, vehicle manufacturers and importers, the media, and fleet managers.
- Limits of human performance: people make mistakes, and the need to acknowledge our capability limits.
- The physical limits of human tolerance to violent forces: we are physically vulnerable when in a crash.
- A forgiving road system: so that when crashes do happen, deaths can be avoided and injuries minimised. In other words, a road system that does not allow human error to have a serious or fatal outcome.
- The need to improve the safety of all parts of the system: roads and roadsides, speeds, vehicles, road use, and post-crash care – so that if one part fails, other parts will still protect the people involved. The approach involves all elements of the road transport system working together to prevent crashes or limit crash forces, making them survivable and reducing the severity of injury.

Elements of a Safe System. The Safe System in Australian road transport generally focuses on five key elements of the system.

- *Road users* will be appropriately trained and licenced, alert and aware of the risks, and drive or ride to the conditions; there will be more in-vehicle technologies to give drivers safety feedback, ensure alertness and reinforce compliance with the road rules.
- *Roads and roadsides* will be safer because transport and urban planning, and road design, will accommodate errors; surfaces will be improved and roadside hazards removed or barriers installed.
- *Vehicles* will increasingly have advanced safety features, including electronic stability control, airbags, head restraints, and collision avoidance systems. HCD is now well recognised in the automotive domain to play a key role in developing fit-for-purpose road transport technology.
- *Speed* will be managed to safe levels through more appropriate limits, and there will be smarter "self-explaining" roads and roadsides that guide people about what safe speeds mean.
- *Post-crash care* will occur efficiently and effectively following incidents. Post-crash care is required to minimise the likelihood of crashes resulting in a fatal or serious injury.

3.4.3 Examples

Two examples are given: the first focuses on the Safe System/human systems integration and the second on human-centred design.

Wire Barriers and the Safe System. Candappa (2020) investigated the effectiveness of wire rope safety barriers along highways to prevent fatal and serious injuries. She used a safe system framework for this work, with the wire barrier as a key component of safe roads and roadsides within a forgiving road transport system. Within the safe system framework, wire barriers were found to be highly effective in reducing serious casualty (fatal and serious injury) off-path crashes relative to guardrails or concrete barriers.

Human-Centred Design in Road Transport. Human-centred design is widely used by automotive OEMs, for example, D-Ford is a human-centred design lab that works within Ford to champion the user's voice. With regulators and highway authorities, human-centred design is now also frequently used, for example, the European Union's HUMANIST (HUMAN centred design for Information Society Technologies) network for new transport technology. As a concrete example, Horberry et al. (2021) used a human-centred design approach to design and evaluate a truck driver fatigue and distraction warning system: they used a multi-stage iterative process of a comprehensive literature review, developing a context of use description, undertaking truck driver interviews, identifying user needs and associated design requirements, conducting two design workshops, operationalising the design, running interface evaluation studies, and finalising the concepts to develop an effective driver warning system interface aimed at commercial truck operations.

3.4.4 Implications for Mining

The first key lesson learnt from human systems integration in the road transport domain is the need to move the focus from the individual driver towards a shared responsibility for safety and human element risks across the whole transport system. This wider Safe System approach has had a great deal of success in terms of reducing deaths and serious injuries in countries that have adopted it. The coal mining industry in Australia does cover many of the aspects of a Safe System but does not yet formally adopt the overall shared responsibility approach and key guiding principles. Further work to emulate the successes of the Australian road transport system is recommended.

The second key lesson from this domain is the success of human-centred design, particularly for vehicle and transport technology design. Human-centred design is generally now seen as a valuable core component of design, rather than as an expensive add-on, and is often used as a selling point by automotive OEMs. Parts of the coal mining industry are beginning to recognise the benefits of human-centred design, but the whole approach in not yet as endemic as in the road transport domain. Further formalisation of coal mining human-centred design is therefore recommended.

3.5 RAIL

3.5.1 Overview

There is no single leading practice approach to human-centred design/human systems integration within the global rail industry. Instead, leading approaches from rail operations in the USA and Australia will be presented here. In both countries, ISO standards – both rail-specific and general – are endemically used, and these include:

- ISO 9241:210 (2019): Ergonomics of human-system interaction: Human-centred design for interactive systems
- ISO 9241-220 (2019): Ergonomics of human-system interaction: Processes for enabling, executing, and assessing human-centred design within organisations
- ISO 6385 (2016/2021): Ergonomics principles in the design of work systems
- EN 50126-1 (2017): Railway applications – the specification and demonstration of reliability, availability, maintainability and safety (RAMS)
- EN 50129 (2018): Railway applications – Communication, signalling, and processing systems – Safety-related electronic systems for signalling
- ISO 11064 (Parts 1–7, various dates): Ergonomic Principles in Control Room Design
- AS7470 (2016): Human Factors Engineering Integration

In the USA and Australia, human-centred design and human systems integration approaches are well-embedded in engineering design processes and rail operations. In US rail the term "Human Systems Integration" is employed, whereas in Australian rail "Human Factors Integration" is more common: but the processes in both countries are very similar. For example, AS7470 was specifically prepared to support Human Factors Integration (HFI) in the engineering design process within the Australian Rail Industry. This includes the requirements for organisations conducting or procuring engineering design activities, services, or products to:

- incorporate Human Factors within their engineering design processes,
- ensure their products comply with the generic Human Factors requirements in the standard, and
- use the Human Factors Integration process to identify the specific Human Factors requirements of the system or asset being designed, procured, or modified.

The aim of the requirements specified in AS7470 is to optimise overall system performance through the systematic consideration of human capabilities and limitations as inputs to an iterative design process. Adequate integration of Human Factors in all phases of a system's development lifecycle ensures its safety, performance, and fitness for purpose. Equally, the aim of the human factors integration process is to identify, then mitigate and prevent Human Factors-related risk, and ensure that human-system interactions are optimised for system performance and safety. Incorporating human factors integration into the engineering design process also facilitates a high level of system acceptance amongst end users.

Likewise, in US rail, human systems integration is defined as a "systematic, organization-wide approach to implementing new technologies and modernizing existing systems". It combines methods, techniques, and tools designed to emphasise, during the acquisition process, the central role and importance of end-users in organisational processes or technologies. Human systems integration here refers to efforts to increase safety, manage risk, and optimise performance of those who work in socio-technical rail systems. Human systems integration considers the human role (both individuals and teams) as part of a system that includes tasks, technologies, and environments. It ensures that characteristics of people are considered, and accounted for, throughout the design and development of systems. As with AS 7470, the US human systems integration approach can increase operator acceptance and usability of technology, and enhance the likelihood of successful adoption (FRA, 2019).

3.5.2 Examples

Positive Train Control (PTC). In December 2020, the Federal Railroad Administration (US) announced that positive train control technology is in operation on all 57,536 required freight and passenger railroad routes. Positive train control systems are designed to prevent train-to-train collisions, over-speed derailments, incursions into established work zones, and movements of trains through switches left in the wrong position.

Among the changes facing the industry is the introduction of new technology in the locomotive cab, such as positive train control, intended to increase safety and minimise accidents. When new technologies are introduced, it is important that it is done in a way that enhances, rather than hinders, operator performance. Human systems integration processes were used to ensure that this happens and that new technologies are optimally designed for safe, efficient, and reliable use. This included identifying human and organisational factors that affect a successful positive train control implementation, such as the application of High Reliability Organising (HRO) characteristics in the implementation of this safety system (Khashe & Meshkati, 2019). A focus on Human Factors and the end-user in all phases of a positive train control's development lifecycle ensures its safety, performance, and fitness for purpose.

3.5.3 Implications for Mining

A key lesson is the potential value in developing the mining equivalent of AS7470, that is, Human Factors Engineering Integration for the mining industry. This would be applicable across the system's development and deployment lifecycle and would focus both on safety and performance issues.

3.6 HEALTH

3.6.1 Overview

Since the 1970s, "medical devices" marketed in the USA have been regulated by the Food and Drug Administration (FDA). Current federal regulations oblige manufacturers of all but the most innocuous medical devices to implement a process to control device design (Design Controls, 21 C.F.R. § 820.30, 2021). The regulations outline a general framework for manufacturers to use, which applies to both initial design processes and processes of re-design and refinement conducted after the product has been introduced to the market. A complementary FDA guidance document provides advice on implementing the process (Food and Drug Administration, 2016b); while another recommends that manufacturers apply human factors and usability engineering processes to the design of medical devices to ensure that they are safe and effective for their intended use, users, and use contexts, with particular emphasis on points of interaction between humans (including clinicians and patients) and the device (including displays, controls, packaging, labels, instructions, etc.; Food and Drug Administration, 2016a). Taken together, these requirements and recommendations represent an endorsement of human-centred design activities as leading practice for the development of medical devices.

The FDA guidance (Food and Drug Administration, 2016a) acknowledges that "use-related hazards" (i.e., those associated with interactions between users and the device, such as "aspects of the interface design that cause the user to fail to adequately or correctly perceive, read, interpret, understand or act on information from the device"; p. 5) are a potential source of harm to clinicians and patients, and that these can arise when: use of the device places excessive demands (physical, perceptual, or cognitive) on the user; device use is counter-intuitive or inconsistent with expectations; the use environment impairs operation of the device or the user's abilities when using it; or the device is used inappropriately in a way that could have been anticipated but was not, or was anticipated and was not adequately controlled.

Specific design practices recommended by the FDA guidance include:

- define the intended population of device users (including their professions, physical and cognitive attributes, training, experiences, and motivation), relevant characteristics of the potential use environments, and all elements of the device that users will potentially interact with (i.e., the user interface);
- conduct preliminary analyses (e.g., task analysis, heuristic evaluation, cognitive walk-throughs, or simulated-use testing) to identify user tasks and potential use-related hazards (including the circumstances in which they arise);
- eliminate or reduce design-related problems that cause or contribute to unsafe or ineffective use;
- conduct user-centred human factors validation testing to assess the effects of the final design on the safety and effectiveness of use; and
- document all of the above.

The FDA guidance also includes a summary reporting template that manufacturers can use to provide information to the FDA about the safety of the device and the investigations conducted to assess and address human factors issues. The template encompasses the following:

1. An overall conclusion about the safety and effectiveness of the device for the intended users, uses, and use environments.
2. A description of the intended users, users, use environments, and training.
3. A description of the device user interface (including a graphical representation).
4. A summary of known use problems (including problems with previous models and similar devices, and design changes implemented to address them).
5. Analysis of hazards and risks associated with the use of the device (including potential use-related errors, the associated potential for harm, and measures employed to reduce the risk).
6. Summary of preliminary analyses and evaluations (including methods, key results, and consequent design modifications).
7. Description and categorisation of critical tasks (including how they were identified)
8. Details of human factors validation testing (including the methods used, the rationale for selecting them, the test environment, the training given to participants, and the results).

While the FDA's recommended practices represent current accepted leading practice for the design of medical devices, several caveats must be taken into consideration:

1. The practices recommended by the FDA amount to a substantial programme of human-centred design, if implemented appropriately; however, the recommendations do not amount to a full human-systems integration process (e.g., there is no consideration of factors such as cost of ownership or the impact of the device on total system performance beyond its immediate use environments).
2. Some forms of health information technology, such as electronic health records (EHRs) have not been regarded as "medical devices", and have therefore not been subject to the same regulation and FDA guidance as traditional medical devices, despite the potential risks to users and patients associated with their use (Konnoth, 2022). Hence, current practice in the development and implementation of EHRs is typically sub-optimal (Abbott & Weinger, 2020) and cannot be considered good practice. Consequently, EHRs are still developed and deployed have poor usability, impose an excessive cognitive workload on clinicians (often in the service of administrative concerns not directly related to the provision of high-quality care), and are implicated in safety issues (Abbott & Weinger, 2020). Further, the implementation of such systems has been found to be associated with a

range of negative outcomes for users, including clinician dissatisfaction and burnout (which itself is associated with poor clinician well-being, reduced quality of care, reduced safety, patient dissatisfaction, and increased staff turnover; Abbott & Weinger, 2020). This suggests an urgent need to not only expand the use of human-centred design for EHRs but also to adopt effective human-systems integration processes.

3. Despite being a ubiquitous "old technology", the design of paper-based clinical charts (e.g., observation charts for recording a patient's vital sign data) can affect clinicians' interpretation of patient data with potential consequences for the detection of clinical deterioration, yet these *cognitive artifacts* are also not regarded as "medical devices" and are seldom subjected to substantial human-centred design processes (Christofidis et al., 2016).

3.6.2 Examples

Implementation of EHRs in Queensland, Australia: A Cautionary Tale. Scott et al. (2018) provided a checklist intended to assist hospitals in preparing for the implementation of EHRs, based largely on the roll-out of such a system in what became Australia's first digital tertiary hospital (which was subsequently extended to other hospitals in Queensland). Notable omissions from the approach described include any mention of human factors input/expertise or methods, usability testing (vs. user-acceptance testing), and human-centred design. Further, despite a strong emphasis on the technical aspects of integration, there was limited emphasis on the human aspects (i.e., human-systems integration). In addition, the authors explicitly endorsed the use of workarounds for software design issues that pose a threat to patient safety, rather than making design improvements before the system goes live – in stark contrast with a human-centred design or human-systems integration approach. Media reports in relation to the system have alleged significant clinician concerns around its usability and its negative impacts on hospital productivity and patient safety, including increases in "wrong blood in tube" incidents at multiple hospitals (e.g., https://www.brisbanetimes.com.au/national/queensland/doctors-demand-credible-independent-review-of-troubled-health-record-system-20200608-p550j2.html).

Development and Implementation of the ADDS Chart: A Success Story. The Adult Detection Deterioration System (ADDS) Chart is a paper-based hospital observation chart primarily for recording hospital patients' vital signs (e.g., blood pressure, heart rate). It was developed by a multidisciplinary team of human factors specialists and clinicians who employed a substantial human-centred design process that included task analysis, heuristic evaluation, iterative design with rapid prototyping, and simulation-based usability testing. Evidence from the usability studies indicated that both experienced clinicians and novice chart-users were faster and more accurate at detecting abnormal observations when using the ADDS chart, compared with pre-existing clinician-designed forms, and made fewer scoring errors relative to charts with comparable systems for scoring the severity of abnormal observations. Subsequent independent clinical implementations of the chart in hospitals in Australia and New Zealand have been associated with improvements in a range of patient outcomes (including reductions in the rate of in-hospital cardiac arrests), as well as high rates of user satisfaction (for an overview of the chart development and related empirical evidence, see Cornish et al., 2019).

3.6.3 Implications for Mining

The overall lesson from the healthcare industry is that regulation can be an important driver of good practice in the area of product design (e.g., highly regulated traditional medical devices vs. poorly regulated EHRs), and may therefore be a "necessary evil" for mining technology/equipment.

In jurisdictions where this is not already the case, new regulations should be formulated that oblige mining technology/equipment manufacturers to produce human-centred design summary reports, like those required by the FDA for medical devices (Food and Drug Administration, 2016a), showing how they have identified and addressed potential human factors issues. In the absence of regulations, mining companies should nevertheless demand such reports during procurement processes.

Finally, even "old technologies" can be improved through human-centred design processes, with potential safety benefits where the "old technology" performs a safety-critical role.

3.7 CHAPTER CONCLUSIONS

Leading practice in the defence, aviation and airspace, road transport, rail, and healthcare domains was presented, including relevant individual domain ISO and national standards. Implementation examples were provided, and subsequent implications for mining implementation arising from each domain were discussed. Key lessons included the importance of considering human systems integration as a fundamental component of the acquisition/procurement process (defence), the importance of human-centred design application to procurement at all levels (aviation and aerospace), the potential value of developing a mining equivalent of AS7470 Human Factors Engineering Integration for the mining industry (rail), the role of regulation as an important driver of good practice in the area of product design (health), and shifting the focus from individual to shared responsibility and key guiding principles for safety and human element risks across the whole system (road transport). The following chapter discusses existing guidance material for the implementation of automation in the mining sector.

4 Existing Guidance for the Mining Industry

4.1 INTRODUCTION

This chapter summarises the standards and guidelines that have been developed globally to assist the safe implementation of automation in mining. Particular attention is given to guidance provided regarding the human aspects of automation, and the sufficiency of this guidance.

4.2 EXISTING STANDARDS AND GUIDANCE MATERIALS FOR MINING AUTOMATION

Albus, J. Quintero, R., Huang, H-M & Roche, M. (1989). Mining automation real-time control system architecture standard reference model (MASREM). NIST Technical Note 1261 Volume 1.

Sponsored by the US Bureau of Mines, the "Mining automation real-time control system architecture standard reference model" was "intended as a reference document for the specification of control systems for mining automation projects" and "provides the high-level design concepts to be used in the automation of a mine". Given the cover image is a stylised underground coal continuous mining machine, it can perhaps be assumed that underground coal mines were the initial target. The advice provided remains relevant.

The document includes sections discussing the role of humans in the system, viz.,

> The sharing of command input between human and autonomous control need not be all or none. The combination of automatic and operator modes can span an entire spectrum from one extreme, where the operator takes complete control of the system from a given level down so that the levels above the operator are disabled, to the autonomous mode where the operator loads a given program and puts the mining machine on automatic. In between these two extremes is a broad range of interactive modes where the operator supplies some control variables and the autonomous system provides others. … Even in cases where the operator takes complete control, some of the higher level safety and fault protection functions should remain in operation.
>
> **p. 19–20**

> The operator interfaces allow the human the option of simply monitoring any level. Windows into the global memory knowledge base permit viewing of maps of a section, geometric descriptions and mechanical and electrical configurations of mining

DOI: 10.1201/9781003380887-4

> machines, lists of recognized objects and events, object parameters, and state variables such as positions, velocities, forces, confidence levels, tolerances, traces of past history, plans for future actions, and current priorities and utility function values. These may be displayed in graphical form; for example, using dials or bar graphs for scalar variables, shaded graphics for object geometry, and a variety of map displays for spatial occupancy. Time traces can be represented as timeline graphs, or as stick figures with multiple exposure and time decay.
>
> Geography and spatial occupancy can be displayed as a variety of maps, vectors, or stick figures, or shaded graphics images. … The operator may also have a direct television image of the mining machine's environment with graphics overlays which display the degree of correlation between what the mining machine believes is the state of the world and what the human operator can observe with his/her own eyes.
>
> **(p. 21)**

> The human operator can thus monitor, assist, and if desired, interrupt autonomous operation at any time … to take control, to stop the mining machine, to slow it down, to back it up, or to substitute the human's judgment by directly entering commands or other information to replace what the robot had otherwise planned to do.
>
> **(p. 22)**

A (brief) "Safety System" section was also included, reproduced here in its entirety.

> The mining machine control system should incorporate a safety system which can prevent the mining machine system from entering forbidden volumes, both in physical space and in state space. This safety system should always be operational so as to prevent damage to the mining machine or surrounding structures or humans during all modes of operation: teleoperation, autonomous, and shared.
>
> The safety system should have access to all the information contained in the world model of the control system, but should also maintain its own world model, updating it with redundant sensors. The safety system should periodically query the control system to test its state and responsiveness. Conversely, the control system should also periodically query the safety system to test it. Observed states should be constantly compared with predicted states and differences noted. If either system detects an anomaly in the other, error messages should be sent and appropriate action taken.
>
> **(p. 22)**

4.2.1 Department of Mines and Petroleum (2015): Safe Mobile Autonomous Mining in Western Australia – Code of Practice

The code of practice provides guidance for mobile autonomous and semi-autonomous systems used in surface and underground mines in the Australian state of Western Australia. The document adopts a conventional risk management approach and notes that autonomous mining may bring additional risks. A general duty of care that applies to all stakeholders is noted, and the code encourages "continuing communication and consultation between system and component suppliers and the mining operation" (p. 2).

The code defines responsibilities for two groups: system builders – those who supply the automation; and system operators – those who use the automated mobile equipment.

Amongst other responsibilities that include determining functional safety requirements and establishing performance requirements, system builders are required to share residual risk information with the system operator. Responsibilities of the system operators include: understanding the risks including any residual risks, developing safe work procedures, consultation with workers, providing training, and investigating incidents.

Three questions are posed for the risk management process:

- "What are the potential scenarios for mobile autonomous mining incidents?
- What are their potential consequences in terms of safety and health?
- What controls are available and how effective are they?" (p. 4).

Starting with these questions is more effective than commencing with considering failure modes because it encompasses the possibility that incidents with potential adverse consequences may arise in the absence of the failure of any system component. Appendix 6 of the code of practice provides examples of incident scenarios, including:

- unauthorised access of personnel or equipment into autonomous area
- autonomous equipment going into unauthorised areas or performing tasks that cause safety risks (it is suggested that this may occur due to human errors such as overriding an alarm condition, or failure to update a map)
- communications failure
- autonomous equipment loss of traction
- failure to communicate system changes
- unintended traffic interactions
- inadvertent switching between modes
- interactions with pedestrians
- remote re-starting of autonomous vehicle from an inappropriate location
- fire

Risk identification techniques nominated include HAZOP and LOPA, and reference is made to functional safety analysis. The code of practice recommends that:

"those undertaking a risk assessment have the necessary information, training, knowledge and experience of the:

- operational environment (e.g. scale, complexity and physical environment of mining activities)

- operational processes (e.g. maintenance systems, work practices, interaction, separation)
- autonomous systems (e.g. functionality, safety features)." (p. 4).

While these areas of expertise are no doubt important, the code of practice is deficient in failing to identify the understanding, training, knowledge, and experience of humans as critical to the risk assessment.

Emphasis is placed on the importance of higher order control measures, monitoring and review, and documentation of the outcomes of the risks assessment process and subsequent monitoring activities in the operation's risk register.

The importance of training is highlighted in the code of practice. It is noted that personnel must be provided with the knowledge and skills required to perform required tasks and particularly how to recognise when the system is not operating as intended, and what actions to take in such situations. A requirement for evidence-based assessment of competency is highlighted, as is the importance of consultation, retraining, and reassessment whenever changes are made to the systems of work. Supervisors are identified as a "fundamental safety function", with responsibilities including ensuring work is carried out as intended and verifying that the system continues to function safely.

Section 5 of the code of practice focuses on mine planning and design for controlling hazards. It is suggested that mine design should be suitable to autonomy and aim to minimise interaction with people and manual equipment. Attention is directed, in particular, to road design, traffic management (including intersections, load and dump locations, access controls), and separation of autonomous equipment from people and manual equipment.

Section 6 of the code of practice places particular emphasis on the role of "functional safety" standards, suggesting that:

> Functional safety provides assurance that the safety-related elements of the autonomous system and operational controls provide suitable risk reduction to achieve the safe operation of the autonomous systems.
>
> **(p. 12)**

While functional safety approaches are important for assessing the reliability of individual components in the system, the faith that this assures safe operation of the entire system is misplaced. The code of practice does note at this point the relevance of human interactions with autonomous systems and the potential impact of human behaviours for the assessment of risks.

Section 7 of the code of practice lists issues to be addressed during commissioning including "functional and user acceptance testing"; however, the code is silent on the involvement of the real users, that is the people who will be required to work with the system, during this acceptance testing.

Section 8 of the code of practice highlights "operational hazard controls", listing issues to be addressed by operational practices. The importance of supervision, training and competency assessment, and change management is reiterated; as well as directing attention to rules governing changes between autonomous and manual operating modes, and traffic management rules that govern interactions between autonomous equipment, manual equipment, and pedestrians. At this point "human factors (eg., response to system information or warning, adherence to exclusion zones)" (p. 13) are listed as a matter to be addressed within operation practices, along with area security and control. This is too little, and too late!

Section 9 of the code of practice notes that maintenance hazards also require consideration. Attention is directed to a series of issues including functional safety considerations for system maintenance, recovery procedures in autonomous areas, and area and activity isolation. The final section of the code provides recommendations related to emergency management.

4.2.2 ISO (2019): Earth-Moving Machinery and Mining – Autonomous and Semi-autonomous Machine System Safety – ISO 17757: 2019

The ISO standard for autonomous and semi-autonomous machine system safety specifies safety criteria and guidance on safe use. The standard also provides definitions for terms related to such machines.

ISO 17757 stipulates that a risk assessment process shall be completed for autonomous and semi-autonomous machine systems according to the principles described in ISO 12100: 2010 Safety of machinery – General principles for design – Risk assessment and risk reduction (ISO, 2010). This standard, in turn, provides a strategy for risk assessment that stipulates that the designer shall: (1) determine the intended use and foreseeable misuse of the equipment; (2) systematically identify the hazards and associated hazardous situations; (3) estimate the risk for each circumstance and hazard; (4) evaluate the risk; and (5) eliminate the hazard or reduce the risk.

Task-based risk assessment is required in that the hazardous situations referred to in Step (2) of this strategy are defined as circumstances in which a person is exposed to a hazard. The standard requires the systematic identification of these circumstances, and notes that to achieve this it is necessary to:

> identify the operations to be performed by the machinery and the tasks to be performed by persons who interact with it, taking into account the different parts, mechanisms or functions of the machine, the materials to be process, if any, and the environment in which the machine can be used. ... All reasonably foreseeable hazards, hazardous situations or hazardous events associated with the various tasks shall then be identified.
>
> **(ISO 12100 Section 5.4, p. 15)**

Further, ISO 12100 Section 5.2 stipulates that the information required for risk assessment (analysis and evaluation) should include the experience of users of similar machines and, wherever practicable, an exchange of information with the potential user. That is, ISO 12100 requires task-based risk assessments and recommends user involvement. ISO/TR 14121-2 Safety of Machinery – Risk Assessment – Part 2: Practical Guidance and Examples of Methods (ISO, 2012) similarly notes that the team conducting a risk assessment should include those with actual experience of how the machine is operated and maintained. This is now possible for many autonomous and semi-autonomous machines given the growing number of global implementations.

Appendix A of ISO 17757 provides a consolidated list of the failure modes identified in the standard specific to autonomous and semi-autonomous machines. The focus on failure modes neglects the potential for unwanted outcomes to occur in complex systems even though all systems function as intended. Appendix B provides further guidance for managing risks specific to autonomous and semi-autonomous machines. The Appendix draws on the content of the Western Australian code of practice and shares the failure to identify an understanding of human factors to the risk assessment.

Section 4 of ISO 17757 requires compliance with functional safety performance levels for the *safety-related parts* of control systems. The standard goes on to specify requirements for specific features including remote stop systems and visual indication of operating mode, and modifies existing standards for braking systems and steering to be suitable for autonomous and semi-autonomous machine systems.

The standard identifies errors in the positional and orientation systems utilised by autonomous and semi-autonomous machine systems as creating risks of unwanted outcomes including collisions with other vehicles. A range of potential failure modes are identified and such systems are required to have the means of detecting the accuracy of the position and orientation data. The systems are required to maintain a safe state when the data is not of the required precision and accuracy (presumably as determined by the risk assessment). Sufficient sensor redundancy is required by the standard to allow a safe state to be maintained in the event of failure of one means of determining position.

Where a digital terrain map (i.e., a topographical description of the site in digital format) is used to maintain safe operating conditions, ISO 17757 requires that the system shall monitor the validity of the map and maintain a safe state in the event of insufficient accuracy of the map (again reliant on the risk assessment to define sufficiency). The standard highlights mechanisms by which the map used by the autonomous system may become inaccurate, including roadway deterioration or changes, calibration or alignment error, or an incorrect version of the map being loaded on the autonomous machine. It is also noted that sudden terrain changes may not be able to be responded to – highlighting a residual risk of failure.

Perception systems (as defined by ISO 17757) are sensors that capture information about the autonomous machine's surroundings. This information is subsequently analysed to detect and classify features or objects of interest. The purpose of such perception systems is to provide information for the safe autonomous control of the machine. A range of failure modes for such object detection systems are identified, including occlusion by contaminants on the sensor; poor lighting; uneven ground; machine vibration causing sensor misalignment; objects too small, or moving too fast to be detected; transparent or dark objects; or a delay in classification caused by other applications overloading the processor used for detection or classification. The possibility of false detections is also highlighted, as is the possibility of inaccuracy in the location of detected objects, or misclassification. The accuracy of the classifier is in part dependent on the quality of the training of the classifier.

Having defined these potential errors, ISO 17757 then stipulates that: the requirements of the perception system shall be based on the risk assessment; the perception system shall maintain the safe state of the autonomous machine during any interaction with its intended operating environment; the autonomous machine system shall detect when the perception system is not meeting the minimum requirements (as determined by the risk assessment) and both maintain the machine in a safe state, and inform the (human) supervisor.

Errors in navigation of the autonomous machine are identified as resulting in risks of collisions with other equipment, infrastructure, or people. Such errors are identified as potentially resulting from inaccurate position and orientation information, incompatible coordinate systems, imprecise navigation control, poor planning, or an inaccurate digital terrain map. The standard requires the autonomous machine system to maintain a safe heading and velocity when operated in accordance with the specified operating environment and conditions; to identify if this is not the case and if so, to take action to maintain a safe state and identify the human supervisor. A potential role for a "competent person" in validating paths or areas to be used by the autonomous machine is also noted.

The role of task planning is noted to vary greatly depending on the machine and the application. Risks identified with task planning are that the autonomous machine could be directed to travel a non-trafficable path or a hazardous path, or to undertake an activity that has undesirable consequences for another machine or person. The standard requires that all such risks: "Shall be noted and mitigated as part of the risk assessment process" (p. 16).

A potential role for humans in assisting the task planner to avoid errors is also noted:

> The task planner shall avoid directing the ASAM onto a known hazardous path. The hazard level of the path may be determined either by the ASAMS or humans interacting with the ASAMS or some clearly defined combination of the two. If the ASAMS is responsible for determining the hazards associated with a path, then the ASAMS shall be able to determine all reasonably anticipatable hazards and have a means to inform the task planner of the detected hazards.
>
> **(p. 16)**

Section 10 of ISO 17757 highlights the importance of communications and networks to the safe operation of autonomous and semi-autonomous machines in mining. Communication failures are associated with a range of risks including loss of remote stop ability; loss of access to situation awareness information; lost or delayed command input; or inaccurate position information. The autonomous machine system is required by ISO 17757 to maintain safe operation in the event of any communications-related failure.

Section 11 of ISO 17757 concerns the supervisor system, including sub-systems such as user interfaces; mission planner; remote control; and configuration management. Risks identified with these systems include incorrect assignment or command provided to the autonomous machine, operation with an incorrect map, or use of incorrect machine parameters. Little guidance is provided regarding the control of these risks.

Section 12 concerns access to the autonomous zone, permissions, and security. Parameters are provided to be taken into account in the risk assessment through which the access control details will be determined. Risks identified in "section 12.3 Operational risks" include several related to human interaction with the system, including "access to the autonomous zone by unauthorised people or equipment"; "ergonomics or human factors that can lead to unexpected switching of operational mode with loss of control"; "improper capture of changes to work areas, especially before switching work areas between manual and autonomous"; "incomplete or improper system updates and changes to programming"; "improper road design, area demarcation or other human errors". Section 12.4 highlights risks associated with mode changes and requires a means to prevent changes that lead to an unsafe condition, including the prevention of unintentional mode change caused by a single human error; and the ability to engage the autonomous mode from a safe position.

ISO 117757 places responsibility for the provision of documentation including manuals, specifications, operating instructions, and training documentation on the system integrator. Section 13.4.2 outlines broad training requirements, including system functionality and hazards and risks; what to expect if environmental or operational conditions change; and how to recognise when machines are not operating as intended, and what actions to take in response. Echoing the Western Australian code of practice, ISO11757 requires evidence-based assessment of competency and suggests that affected personnel should be consulted, retrained as necessary, and reassessed whenever work procedures or plant and equipment change.

4.2.3 GMG (2019): Guideline for the Implementation of Autonomous Systems in Mining

The Global Mining Guidelines Group (GMG) is a not-for-profit organisation founded in Canada in 2012. Corporate members include mining companies, equipment manufacturers, and service providers. The GMG guideline aims to provide information to facilitate the implementation of autonomous systems. It includes references to safety aspects, as well as information on developing a business case, regulations, social impacts, and deployment issues. The safety material is envisaged to form part of an implementation plan describing a risk management process following ISO 17757.

The knowledge identified as necessary for the risk assessment is expanded beyond that identified in the Western Australian code of practice and ISO 17757 to include the "Site continuity plan" and "Corporate risk guideline". Again, the opportunity to note the importance of an understanding of human behaviour was missed.

The role of humans is acknowledged briefly at the conclusion to this section, viz.,

> When determining how to implement various controls, mine sites must ensure they provide sufficient information for decision-making. Though the systems are autonomous,

> human decisions are still required to overcome exception states for autonomous systems at all maturity levels. The system should be designed such that alerts and alarms on the machines and in the control room are prioritized with humans in mind.
>
> (p. 13)

This text makes it clear that the role of humans is considered to be of peripheral importance, envisaged as exception management only. This view is reinforced through a figure provided to illustrate the "key considerations for a design management framework for mining automation". Human-Systems Integration, while acknowledged, is relegated to the periphery as a "broader design context" along with "environmental and social considerations".

However, in Systems Engineering (as defined by the International Council on Systems Engineering, 2015), human systems integration refers to engineering processes that ensure that human-related issues are adequately considered during system planning, design, development, and evaluation – of which safety and health issues are a subset, along with training, human factors engineering, and staffing decisions.

4.2.4 NSW Resources Regulator (2020): Autonomous Mobile Mining Plant Guideline – DOC20/690069

The New South Wales Resources Regulator is a state government agency that is responsible for regulating health and safety for the mining and petroleum industry in the Australian state of New South Wales. The autonomous mobile mining plant guideline summarises the regulator's requirements for sites intending to implement autonomous equipment. The guidance acknowledges the potential safety benefits of automation while noting the potential for new risks to be created. Examples of such new risks are suggested to include:

> automation associated with longwall mining equipment has created new risks, such as being crushed by an automatically advancing roof support.
>
> (p. 4)

The potential for loss of direct perceptual information is noted.

The guideline describes the legislative requirements applicable in New South Wales. In addition to a duty of care imposed by the Work Health and Safety Act, mining-specific regulations impose duties that include conducting risk assessments:

> with appropriate regard to the nature of the hazard, the likelihood of the hazard affecting the health and safety of a person, and the severity of the potential health and safety consequences.
>
> (p. 4)

The guideline directs mine operators to pay particular attention to interactions between people and autonomous machines, noting:

> While autonomous operation, by definition, means there will not be people onboard machines, it does not mean there will be no people in the autonomous operating zone (AOZ). There are tasks such as workplace inspections, machinery inspections, maintenance tasks and repairs. The intent of these tasks is to ensure a safe work environment and the correct operation of the autonomous machines and other plant. Other activities form an essential part of the mining process, such as road maintenance, operator change-over for manually controlled plant and operating service and ancillary vehicles.
>
> Risk assessments for the introduction and operation of autonomous machines must consider all foreseeable scenarios where it is possible for people to interact with the machines, or where the machines may interact with other equipment or infrastructure.
>
> **(p. 8)**

The NSW regulator adopts the approach described by the International Council on Mining and Metals (ICMM) health and safety critical control management implementation guide (ICMM, 2015). Mine operators are directed by the regulator to identify critical controls to prevent incidents associated with autonomous machines, and to implement effective mitigating controls to protect workers in the event that an incident does occur. Consideration is directed to the full life cycle of the equipment.

It is suggested that a mine's critical control management process should include routine verification of the effectiveness of critical controls. Finally, the importance of change management is identified:

> Mine operators must be vigilant in applying their risk management processes to changes as the operation of autonomous machines evolves and expands. The temptation to expand the scope and broaden the use of autonomous machines without appropriate risk management will lead to weakening or loss of existing controls. It may also lead to a failure to identify that additional controls are required due to the changes. Change management processes must be applied to all aspects of machine operation, including hardware and software, and should be used during the complete lifecycle of the machine, including commissioning, maintenance and repair activities.
>
> **(p. 9)**

4.2.5 Alberta Mine Safety Association (2020): Autonomous Haul Systems

The Alberta Mine Safety Association is an industry association formed in 1982 comprised of representatives of mining, quarrying, and oil sands operations in the Canadian province of Alberta. The guide is intended to provide "direction for the healthy and safe application and operation of autonomous technology in mining" for the province and purports to be "Accepted and Approved by Alberta Occupational Health and Safety". While the title of the document suggests a narrow scope, the guidance is intended to be applicable to surface and underground loaders, drills, water carts, and other mobile auxiliary equipment.

The guide directs attention to the risk management framework provided by ISO 12100 and ISO 17757. Constant engagement with "frontline" personnel during the process is identified as "imperative". The importance of training and competency for anyone working in the autonomous zone is emphasised, and the contribution of humans to safe operation is identified, viz.,

> "Although the operation is autonomous, there is a human element that must be accounted for so the operation can run safely. This risk can be managed through an organization willingness to:
>
> - give workers the training and time to build the necessary skill sets to operate within the AOZ safely
> - commit to regular discussions about automation project challenges and collaborate management strategies with front line workers.
> - apply best practice and sharing of information from previous autonomous operations for learning and to support new challenges within the autonomous project" (p. 7).

Additional guidance, including examples of hazards that may be encountered at pre-implementation, implementation, and operation stages, is provided.

4.2.6 Alberta Occupational Safety and Health (2020): Applying for Occupational Health and Safety Autonomous Haulage System Approval

The document summarises the steps required for a site to obtain approval to operate an autonomous haulage system. Documentation to be included in the application includes braking, steering, and other primary system test results; as well as a detailed project management plan that is to "establish a thorough knowledge of the AHS as well as the hazards and controls that are associated with the technology". A detailed outline of the topics to be addressed in the plan is provided, and this includes a safety management plan. A range of additional appendices are required – including a letter from the manufacturer indicating compliance with ISO 17757.

4.2.7 Mandela Mining Precinct (2021): Guideline – Best Practice Applications of Mechanised Equipment

The Mandela Mining Precinct is a Public-Private Partnership between the South African Department of Science and Innovation through the Council for Scientific and Industrial Research, and the Minerals Council South Africa. The focus of the guide is on underground narrow reef hard rock mining in South African gold and platinum mines. The document draws heavily on the GMG (2019) guideline. Some country-specific issues are noted. For example, it was noted that a tele-remote initiative at one site was abandoned due to union opposition.

The guideline suggests that: "All mine design and planning must incorporate safety by design" (p. 26) noting that design considerations include human/system interfaces; however human aspects are notably absent from the list of topics to be considered in "safety by design". Attention is directed instead to a list of engineering standards provided by GMG (2019).

Chapter 3 of the guideline provides general occupational health and safety considerations for mechanised mining systems. The importance of participation of all stakeholders in development of an OHS management plan is emphasised. Specific to automation, the following are suggested for consideration:

- "A real-time tracking system to flag personnel entering 'no go' zones.
- A system to indicate personnel as either 'at shaft' or 'safe from shaft' when tagging in or out at the lamp room.
- Machines should have pedestrian detection systems to avoid injuries.
- With the potential reduction in ventilation requirements in autonomous operations, the real-time tracking system should be utilised as input to 'ventilation-on-demand' systems" (p. 51).

Additional chapters of the guideline address regulatory requirements and change management.

4.2.8 Resources Safety & Health Queensland (2022): Autonomous Mobile Machinery and Vehicle Introduction and Their Use in Coal Mining – Queensland Guidance Note

The guidance note was issued by the Mines Inspectorate to guide surface and underground coal mining operations in the Australian state of Queensland, drawing largely on the Western Australian code of practice. Responsibilities for the safe implementation of autonomous and semi-autonomous mobile machinery are broadly divided between system builders and system operators. It is suggested that the responsibilities should be defined and agreed by all parties.

It is noted that input from many people is required for effective risk management. Those nominated include: "researchers, design engineers, project managers, team leaders, controllers, safety and health representatives, coal mine workers involved in the tasks, and emergency response personnel" (p. 8).

The guideline suggests that the safety functionality of autonomous control systems should take into account, amongst other things, an assessment of interactions between personnel (operational and maintenance) and the autonomous systems, and should consider "the impact of human interactions and behaviours on autonomous system performance" (p. 12). Considerations for operational practices are listed, including operating team's technical knowledge; change management; interaction rules; and "human factors (e.g. response to system information or warnings, adherence to exclusion zones)" (p. 14). Maintenance safety considerations are also listed, including recovery procedures in autonomous areas, isolation, calibration and testing.

The importance of ensuring work area design minimises interaction between autonomous machines and manual vehicle and people is noted. Considerations here are listed as including access controls, consumable resupply, loading, traffic management, mode changes, and placement of infrastructure.

4.2.9 British Columbia Ministry of Energy, Mines and Low Carbon Innovation (2022): BC Guideline for Safe Mobile Autonomous Mining (Guideline)

The British Columbia guideline is an adaption of the Western Australian code of practice intended for use by mining operations in the Canadian province of British Columbia in the preparation of Autonomous Mining Project Management Plans for submission to the provincial government. This project plan must be "prepared by a qualified professional" and is required to contain many safety-related elements, including a detailed risk assessment; a summary of the health and safety plan; a summary of system safety features; an interaction plan for human-operated equipment; a training program and competency assessment; a process for investigating failures; and a summary of critical controls as identified in the risk assessment.

The guideline also summarises legislative provisions relevant to mines of the province specific to autonomous mining systems. These include the requirement for safe working procedures for autonomous equipment that address, for example: access to autonomous areas; procedures for working within an autonomous area; clearing of autonomous areas for restarting; switching modes; and recovering a failed/stopped mobile autonomous equipment.

Another requirement noted in the guideline is that the legislation in the province (Section 6.19.1 of the Health, Safety and Reclamation Code for Mines in British Columbia) explicitly places responsibility on autonomous system supervisors for system safety, viz.,

> A person who enters commands or inputs information into an autonomous or semi-autonomous system that governs the behavior of tracked or rubber-tired mobile equipment, must do so in a manner that ensures the safe operation of the equipment and that the system can maintain full control of the mobile equipment.

The explanatory notes provided in the guideline note:

> Autonomous mobile equipment is controlled differently than conventional equipment. Conventional equipment has an operator behind the steering wheel who is responsible for maintaining control of the equipment (section 6.19.1 (1)). For an autonomous system, there can be a variety of people who are responsible for control of the system, including, but not limited to, individuals who survey the working area, restart stopped equipment, design the digital environment, assign tasks, or input commands to the system. *Any individuals who enter commands or input information into the system have a responsibility to ensure the system can maintain control of the equipment.*
>
> **(p. 28; Emphasis added)**

It is not clear whether this responsibility extends to the system designers, for example, those who coded the software.

An additional obligation to prepare traffic control procedures placed on "the manager" by section 6.8.3 of the Health, Safety and Reclamation Code is noted to be applicable to autonomous mining operations. It is suggested that these procedures should include:

> Rules for interactions between conventional and autonomous equipment; Autonomous operating area access and exclusion zones; Road, dumping and loading area design requirements and system limitations from manufacturer; and Priority rules.
>
> **(p. 30)**

4.3 FUNCTIONAL SAFETY GUIDELINES

Functional safety has historically formed a core component of efforts to ensure the safety of automated mining equipment. However, functional safety addresses the safety-related components of the control system rather than the system as a whole, and such approaches do not adequately consider the role that humans play in system safety.

4.3.1 IEC 61508 Functional Safety of Electrical/Electronic/Programmable Electronic Safety-Related Systems

The IEC 61508 series of standards sets out methods for defining and achieving satisfactory performance of safety-related systems. Human factors issues related to the functioning of such systems receive only limited attention, i.e., as IEC 61508-1 explains:

> Although a person can form part of a safety-related system (see 3.4.1 of IEC 61508-4) human factor requirements related to the design of E/E/PE safety-related systems are not considered in detail in this standard.
>
> **(IEC, 61508-1, 1, note 2)**

That noted, the IEC 61508 series of standards refers directly, or indirectly, to human factors issues in several places. IEC 61508-1 requires that:

> The hazards, hazardous events and hazardous situations of the (Equipment Under Control) and the (Equipment Under Control) control system shall be determined under all reasonably foreseeable circumstances (including fault conditions, reasonably foreseeable misuse and malevolent or unauthorised action). This shall include all relevant human factor issues, and shall give particular attention to abnormal or infrequent modes of operation of the (Equipment Under Control) (IEC 61508-1, 7.4.2.3).
>
> **(abbreviations expanded, emphasis added)**

Similarly, it is noted in IEC61508-4 that the risk assessment of Equipment Under Control "will include human factor issues" (IEC61508-4, 3.1.9, Note 3). Little guidance is provided regarding how such consideration of "human factor issues" is to be achieved, although IEC 61508-7 suggests in section C.6.2 that a Software Hazard and Operability Study carried out by a "team of engineers, with expertise covering the whole system under consideration" should "consider both the functional aspects of the design and how the system would operate in practice (including human activity and maintenance)" in identifying hazards and risks.

In providing guidance regarding the application of parts 2 and 3, IEC 61508-6 (Annex A) defines potential failures of the safety system to include both physical faults and potential "systematic faults". The latter includes "human errors" made during the specification and design of a system that causes failure under some combination of inputs, or some environmental condition. The Annex further notes that

> Systematic failures cannot usually be quantified. Causes include: specification and design faults in hardware and software; failure to take account of the environment; and operation-related faults (for example poor interface).
>
> **(IEC 61508-6, Annex A, 1, footnote 5)**

Regarding the human-machine interface, IEC 61508-4 notes that

> a person can be part of a safety-related system. For example, a person could receive information from a programmable electronic device and perform a safety action based on this information, or perform a safety action through a programmable electronic device.
>
> **(IEC 61508-4, 3.4.1, Note 5).**

This note highlights the importance of considering human-machine interactions, and in particular the critical importance of effective interface design. Some information regarding operator interface design is provided in IEC 61508-3 where the software developer is directed to include consideration of "equipment and operator interfaces, including reasonably foreseeable misuse" in the definition of requirements for the system (IEC 61508-3, 7.2.2.5, f).

IEC 61508-3 explains again that "human factor requirements related to the design of E/E/PE safety-related systems are not considered in detail in this standard", however, suggests that, where appropriate:

- "An operator information system should use the pictorial layout and the terminology the operators are familiar with. It should be clear, understandable and free from unnecessary details and/or aspects;
- Information about the (Equipment Under Control) displayed to the operator should follow closely the physical arrangement of the (Equipment Under Control);
- If several display contents to the operator are feasible and/or if the possible operator actions allow interactions whose consequences cannot be seen at

one glance, the information displayed should automatically contain at each state of a display or an action sequence, which state of the sequence is reached, which operations are feasible and which possible consequences can be chosen" (IEC 61508-3, 7.2.2.13, Note 2).

IEC 61508-7 refers to "User friendliness" as a relevant technique referenced in IEC 61508-2 that has the aim of reducing complexity during operation of the safety-related system. The technique is described as:

"The correct operation of the safety-related system may depend to some degree on human operation. By considering the relevant system design and the design of the workplace, the safety-related system developer must ensure that:

- the need for human intervention is restricted to an absolute minimum;
- the necessary intervention is as simple as possible;
- the potential for harm from operator error is minimised;
- the intervention facilities and indication facilities are designed according to ergonomic requirements;
- the operator facilities are simple, well labelled and intuitive to use;
- the operator is not overstrained, even in extreme situations;
- training on intervention procedures and facilities is adapted to the level of knowledge and motivation of the trainee user" (IEC 61508-7 B.4.2).

While these sections in parts 3 and 7 highlight the importance on human-interface design for the performance of the system, the guidance regarding the design of such interfaces is minimalist, and no guidance regarding the evaluation of such interfaces is provided.

4.3.2 ISO 19014 Earth-Moving Machinery – Functional Safety

The ISO 19014 series of standards adapts functional safety methods for application to earth-moving machinery. The approach differs from IEC 61508 in that a "safety-related system" is not defined, and the definition of "safety control system" employed does not include reference to humans. The method defined in ISO 19014-1 starts with identifying possible failure types for the machine control system but differs from IEC 61508 in not including explicit consideration of systematic failures. The method defined also appears to exclude any aspect of the system that is dependent on human reactions as safety-related parts of the control system. For example, section 5, "Requirements for immediate action warning indicators", reads:

> The principles of this standard should also be applied to an immediate action warning indicator intended to warn the operator of a possible hazard and requiring immediate action from the operator to correct and prevent such a hazard.
>
> These indicators shall not be designated as meeting a performance level as the output/diagnostic coverage is reliant on human reaction; indicators provide no control of the system and therefore cannot be labelled as safety-related parts of the control system.
>
> **(ISO 19024-1, 5.1)**

The inference is that human-machine interfaces are not addressed by the ISO 19014 series.

ISO 19024-1 suggests that participants in the development of a machine control system safety analysis should involve:

> a cross functional team, for example, electronic or electrical development, testing or validation, machine or hydraulics design, operator, service, sales and marketing.
>
> **(ISO 19024-1, 6.2)**

There is no comment on the necessity for an understanding of human factors to undertake the assessment – despite requiring that the assessment of the controllability of a hazard take into consideration:

> human reaction (e.g. panic, repeated command of function, etc.) and the capacity for the operator to react to the hazard and provide a means to enter a safe state.
>
> **(ISO 19024-1, 6.5)**

Similarly, ISO 19014-2 excludes consideration of awareness systems such as cameras that do not effect machine motion and excludes audible warnings. ISO 19014-3 relates to environmental performance without relevance to human factors. ISO 19014-4 specifies general principles for software development and signal transmission requirements of safety-related parts of machine-control systems. No human factors input is required nor is any consideration of human-machine interface design principles or evaluation methods included.

4.3.3 ISO 21448 (2022) Road Vehicles – Safety of the Intended Functionality

ISO 21448 provides a complementary approach to functional safety termed "safety of the intended function" that is designed for application to the complex sensors and processing algorithms used in road vehicles to maintain situation awareness. The aim is to avoid "unreasonable risks" due to performance limitations such as (i) the inability of the function to perceive the situation; (ii) lack of robustness of the function with respect to sensor input variations or environmental conditions; or (iii) unexpected behaviour of the decision-making algorithm, rather than system failures.

ISO 21448 Table 1 notes that reasonably foreseeable misuse; and incorrect or inadequate human-machine interface (HMI) (e.g., user confusion, user overload, user inattentiveness) are potential causes of hazardous events that fall within the scope of the standard. Table 5 lists methods for identifying reasonably foreseeable misuse including analysis of use cases and scenarios, analysis of the human-machine interface, and "analysis of human capability to perform or switch between certain tasks". Measures for managing reasonably foreseeable misuse listed include "improving the HMI" (p. 41).

Informative Annex B provides guidance on analysing reasonably foreseeable misuse scenarios based on Human Factors Analysis and Classification System; and section B.4 outlines the use of Systems-Theoretic Process Analysis as a means of analysing the safety of complex systems.

4.3.4 CMEIG/EMESRT/ICMM (2020) White Paper – Functional Safety for Earth-Moving Machinery

A white paper compiled by a collaboration of manufacturers and mining company representatives discussed the application of functional safety approaches to the design of earth-moving equipment. It was noted that automation systems being introduced to earth-moving equipment include non-deterministic elements such as the complex sensors and processing algorithms for situation awareness addressed by ISO 21448 and that such systems cannot be analysed using the functional safety methods provided by IEC 61508 or ISO 19024. It was suggested that while traditional functional safety methods are concerned with identifying and preventing system failures, safety hazards can also occur in complex systems in the absence of failure and recommends the use of Systems Theoretic Process Analysis (STPA) in addition to more conventional risk analysis methods focused on system component failure. No explicit mention is made of human factors, and a figure provided suggests that Functional Safety and Human-Machine Interface are to be considered to be separate aspects of Systems Safety.

4.3.5 GMG (2020): GMG Guideline for Applying Functional Safety to Autonomous Systems in Mining

The Global Mining Guidelines Group has published a "Guideline for applying functional safety to autonomous systems in mining". The guideline scope explicitly excludes non-deterministic elements of the system (e.g., perception systems, artificial intelligence); however, human aspects are referred to in the context of change management, where it is suggested that attention is required: "to confirm that the operations personnel are ready to adapt to the change" and that "Everyone working at the operation should understand the risks of automation for the mine to be safe" (p. 3).

It is also recommended that risk assessments require:

> A strong focus on the administrative controls on which the autonomous system is reliant. They should also consider how human behaviour changes as aspects of manned operation are replaced by the autonomous systems.
>
> **(p. 4)**

However, no guidance is provided regarding how this should be achieved.

A section on competency management is included that suggests identifying tasks to be undertaken and competency criteria for each, including "requirements that demonstrate knowledge, skills, experience, and behaviours" (p. 14). Again, no guidance regarding how these criteria might be derived or assessed is provided.

4.3.6 GMG (2021) White Paper: System Safety for Autonomous Mining

Subtitled "A White Paper to Increase Industry Knowledge and Enable Industry Collaboration on Applying a System Safety Approach to Autonomous Systems", the document notes that functional safety is not sufficient for non-deterministic systems such as those involving machine learning, and including systems reliant on human behaviour. System safety is highlighted as an overarching process involving the use of a safety case. Descriptions of human systems integration (based on Burgess-Limerick, 2020) and software safety management are provided for the education of industry.

4.4 CHAPTER CONCLUSIONS

A range of guidance materials have been developed globally to assist the mining industry in implementing automation safely. This includes Technical Notes from the US Bureau of Mines; ISO standards for earth-moving machinery and mining; mining guidelines for the implementation of autonomous systems from the not-for-profit Global Mining Guidelines Group (GMG); guidance from the various Australian mining states including a Code of Practice from the Western Australian Government; guides/guidance notes from the New South Wales Resources Regulator and Resources Safety & Health Queensland; guidance material from the Alberta Mine Safety Association, Alberta Occupational Safety and Health, and the British Columbia Ministry of Energy, Mines and Low Carbon Innovation; and practice application guidelines from the Mandela Mining Precinct in South Africa. Functional safety guidelines were also presented including the IEC 61508 series of standards which sets out methods for defining and achieving satisfactory performance of safety-related systems; ISO standards; a joint white paper detailing functional safety for earth-moving machinery; GMG guidelines for applying functional safety to autonomous systems in mining; and a GMG white paper on system safety for autonomous mining. The key conclusion here is that existing guidance regarding the human-centred design of mining automation systems is insufficient. The following chapter presents case studies of human systems integration in automated mining operations.

5 Case Studies of Human Systems Integration in Automated Mining

5.1 INTRODUCTION

This chapter presents four case studies of Human Systems Integration in automated mining. They all come from Australia, but each one is applicable to automation in the global minerals industry. Taken together, they show the vital importance of considering operators, controllers, supervisors, and maintenance staff as part of a formal integration programme for both the human and technology aspects of a system. A conclusion section is presented at the end of the chapter which details ten key industry learnings arising from the case studies.

5.2 CASE STUDY 1 – OPERATOR AWARENESS SYSTEM AT GLENCORE COAL ASSETS AUSTRALIA

5.2.1 Overview

Operator fatigue and distraction are long-standing safety issues in mining. Over the past twenty years, fatigue detection technologies have become more reliable and usable and can now form a valuable part of an organisation's fatigue management system.

This case study describes an industry-leading approach by the company Glencore Coal Assets Australia (Glencore) to deploy an Operator Awareness System (OAS). Through careful analysis and preparation, worker consultation, system selection, staged adoption, Trigger Action Response Plans (TARPs), ongoing safety evaluations, and post-implementation management, Glencore's OAS has been successfully deployed across their surface coal mines in Australia.

Overall, using the OAS, Glencore have seen a year-on-year reduction in fatigue events for mobile equipment operators – both in frequency (overall count) and severity. Perhaps the most important finding is that only one recordable fatigue-related incident has occurred across all Glencore Coal Assets sites in Australia since the OAS was fully deployed in 2019.

The case study gives background, details the approach taken, and outlines outcomes in terms of reductions in the number and criticality of fatigue events. Key learnings, discussion and conclusions are then presented.

DOI: 10.1201/9781003380887-5

5.2.2 Background

Operator fatigue, sleepiness, and tiredness have been enduring safety issues in mining and related domains (such as commercial trucking on public roads). Transport and mobile equipment accidents have accounted for a high proportion of fatalities at mines worldwide. Operator fatigue or distraction can be a precursor event to many such vehicle interaction events. Therefore, technologies including operator state awareness systems to help eliminate unwanted vehicle interactions can be a valuable safety management tool (Horberry et al., 2024).

Over the past 20 years, technologies designed to detect fatigue accurately have made significant technological advancements. Most technologies now use either ocular parameters (such as percent of eye closure, or changes in scan patterns), helmet-mounted EEG systems (measuring brainwaves), or hybrid technologies (including those that assess facial expression and head orientation). Also, some systems monitor "operator state" – this can include both fatigue and distraction.

Due to the increased technological maturity of these systems, many organisations are now using fatigue detection technologies or operator awareness systems as a component of their fatigue management system (Horberry et al., 2024). Generally, they are being used for more repetitive mobile equipment tasks such as operating haul trucks, although their use may be expanded to include other vehicles such as water carts, long haul buses that carry personnel to accommodation camps, and some light vehicles (LVs).

5.2.2.1 The Place of Fatigue Management in the EMESRT Nine Layer Control Effectiveness Model

To put the work presented in this case study into a broader mining context, we present an industry-leading model of vehicle interaction controls that was developed by the Earth Moving Equipment Safety Round Table (EMESRT). According to this model, such controls can be divided into the nine levels shown in Table 5.1.

The key controls associated with fatigue detection fit into levels 5 and 8. Ideally, most controls would be higher up the model (levels 1–5).

Fitness to Operate – Level 5
This includes:

- Confirm the operating context
- Review current approaches
- Involve all contributing business functions and stakeholders
- Baseline the status of current controls
- Identify and implement improvement opportunities
- Review support processes for individuals

Fatigue Management – Level 8 (Fatigue Monitoring and Response Technology)
This includes:

- Involve leaders and stakeholders at all levels and identify the value add
- Define what is expected from the technology
- Carry our trials and pilots to confirm accurate, reliable, and repeatable performance
- Cold commission the technology and associated people processes

TABLE 5.1
The EMESRT nine layer control effectiveness model

Level	Defensive controls	Reaction time
1. Site requirements	Creation of equipment specifications and standards, surface mine design	Years
2. Segregation controls	Construction of protective embankments, access control, traffic segregation, work scheduling	Months
3. Operating procedures	Procedure specifications, maintenance, traffic regulations, quality control	Weeks
4. Authority to operate	Training, licensing, induction training, access control	Days
5. Fitness to operate	Fatigue control, drug and alcohol intoxication control, medical check-ups	Shift
6. Operating compliance	Pre-shift inspections, checking equipment health, event recording	Hours
7. Operator awareness	Installation of camera, interactive maps, overview mirrors, alarm lights, visual delimiters	Minutes
8.Advisory controls	Notifications: dangerous proximity, fatigue, over-speed, operating condition of equipment	Seconds
9. Intervention controls	Engine start safety interlock, smooth stop, rollback safety interlock, speed limit	Milliseconds

5.2.2.2 Why Did Glencore Consider Fatigue Detection Technology?

The impetus for the work in this case study was that six fatigue-related incidents involving heavy equipment occurred across six months at Ravensworth Open Cut (ROC) in 2014/15. Consequently, the mine was looking for other methods to manage proactively mobile equipment operator fatigue. Fatigue management at that time was mainly made up of soft controls such as educational sessions, which tended to improve performance for the period of a campaign but then dropped off. New methods to better control fatigue and operator alertness in the long-term were therefore investigated.

Heavy equipment operator fatigue was identified as an area of concern within the company but was unquantifiable at that time as it relied on self-reporting and/or operator confession that an incident involved operator impairment/fatigue.

5.2.3 Approach Taken by Glencore

In 2014/15, the Glencore team investigated what fatigue detection systems were suitable for mining. The key products identified were the Operator Alertness System/

HxGN MineProtect (Guardvant/Hexagon), DSS (Caterpillar/Seeing Machines), and the SmartCap system. During the evaluation process the SmartCap product was dismissed because the system lacked the functionality to verify the fatigue events with video footage. The company also preferred a technology that did not require the operator to wear any part of the product so that it was non-intrusive to the operator.

The systems shortlisted were as followed:

- **Guardvant/HxGN MineProtect Operator Alertness System** (https://hexagonmining.com/solutions/safety-portfolio/hxgn-mineprotect-operator-alertness-system) and
- **DSS** (https://www.cat.com/en_AU/support/operations/frms/monitoring.html)

Both of these systems were trialed at Ravensworth Open Cut (ROC) mine in 2015 onwards. This trial period was for approximately 2 years. During the trial, the Guardvant product was improved through the introduction of digital cameras and improved tracking algorithms. Overall, the Guardvant Operator Alertness System was selected based on achieving numerous functional and performance criteria including its ability to identify eye closure events accurately and consistently as determined through the trial.

It should be noted that Guardvant was acquired by Hexagon Mining in 2018, and some process changes were implemented including the introduction of remote monitoring of system events. This task was initially conducted in the USA; but in 2020, Glencore developed their own monitoring approach (regional-based monitoring centre).

5.2.3.1 Description of the OAS and TARPs at Glencore

The Operator Alertness System (OAS) is a non-intrusive eye closure and distraction monitoring solution specifically built for the mining industry. On-board hardware includes cabin camera and front-facing camera, an infrared sensor, an in-cabin motion alarm and audible device, and a seat vibration device..

The overall process is as follows:

- A dash-mounted camera continually scans the operator's face to detect eye closure events while the truck is in operation (moving at > 5 kph)
- The system is integrated into the fleet management system, which it utilises to transmit its data to the server and to understand which operator is logged onto the vehicle
- When an eye closure event is detected, an in-cab alarm is activated (audible and seat vibration) and the system a video clip of appropriately 6-10 seconds' duration, which is received off-board for review
- A Trigger Action Response Plan (TARP) is then usually followed by the reviewer and the site (Third-party reviewer, Operator, Dispatcher, Mining Supervisor)

During operation, following an eye closure event, the feedback loop is:

1. Operator receives in-cab alarm
2. Third-party reviews event & TARP enacted. Under the TARP, the criticality of the event is :classified into one of three levels according to eye closure duration:
    - Short = yawn. Action: nil by external resources
    - Moderate = uncontrolled eye closure. Action: make contact with the operator
    - Long = eyes closed for multiple seconds. Action: make contact with the operator, truck parked (NSW only), and fitness for work assessment completed by Supervisor (NSW mines only)

In addition to the immediate feeback and other actions outlined above, the Health Intervention TARP is utilised in an attempt to modify the long-term behaviour of operators with significant recurring fatigue events. Using monthly reporting, sites can identify operators with numerous ratings of moderate or long eye closure. These personnel can then be managed through the site's health programme and, if necessary, be sent for health assessments and offered appropriate health interventions (e.g., longer-term health interventions for sleep apnoea.) Incorporating these longer-term follow-ups that address the underyling causes of incidents, as opposed to the system just waking the operators up, contributes to the strength of the overall approach.

5.2.3.2 Glencore OAS Deployment

A trial of the camera-based technology and the associated TARPS was completed at ROC in the first quarter of 2017. The approach taken was to install the system in haul trucks without the in-cab alarms active for the first 14 days. Once the system was verified to be commissioned adequately and baseline data was collected, the in-cab alarms were then activated. The baseline period provided insight into the volume of fatigue events that were occurring prior to system intervention. After the baseline was conducted (typically two weeks), Glencore turned on the in-cab alarms (including seat vibration and auditory warning) and the TARPs came into effect.

The trial provided a better understanding and appreciation of the fatigue risk at ROC, and a decision was made in July 2017 to roll out the OAS system was to all Glencore Coal open cut mines in Australia (11 sites across Queensland and New South Wales), and all systems were installed by the end of 2019. It should be noted that roster types differed between sites, especially between NSW and Qld, and operations in the two states were also subject to different regulations. Consequently, not all sites implemented the fitness for work assessments and health intervention TARP. The deployment process was similar to ROC: generally first involving a baseline period without warnings before full deployment. All systems were installed by the end of 2019. In total, 448 units have been installed on haul trucks and watercarts. A developmental light vehicle unit was trialled at the Newlands operation in 2019..

A diagram of the Glencore Site Fatigue Management Improvement process is shown in Figure 5.1. It should also be noted that project management occurs at each stage.

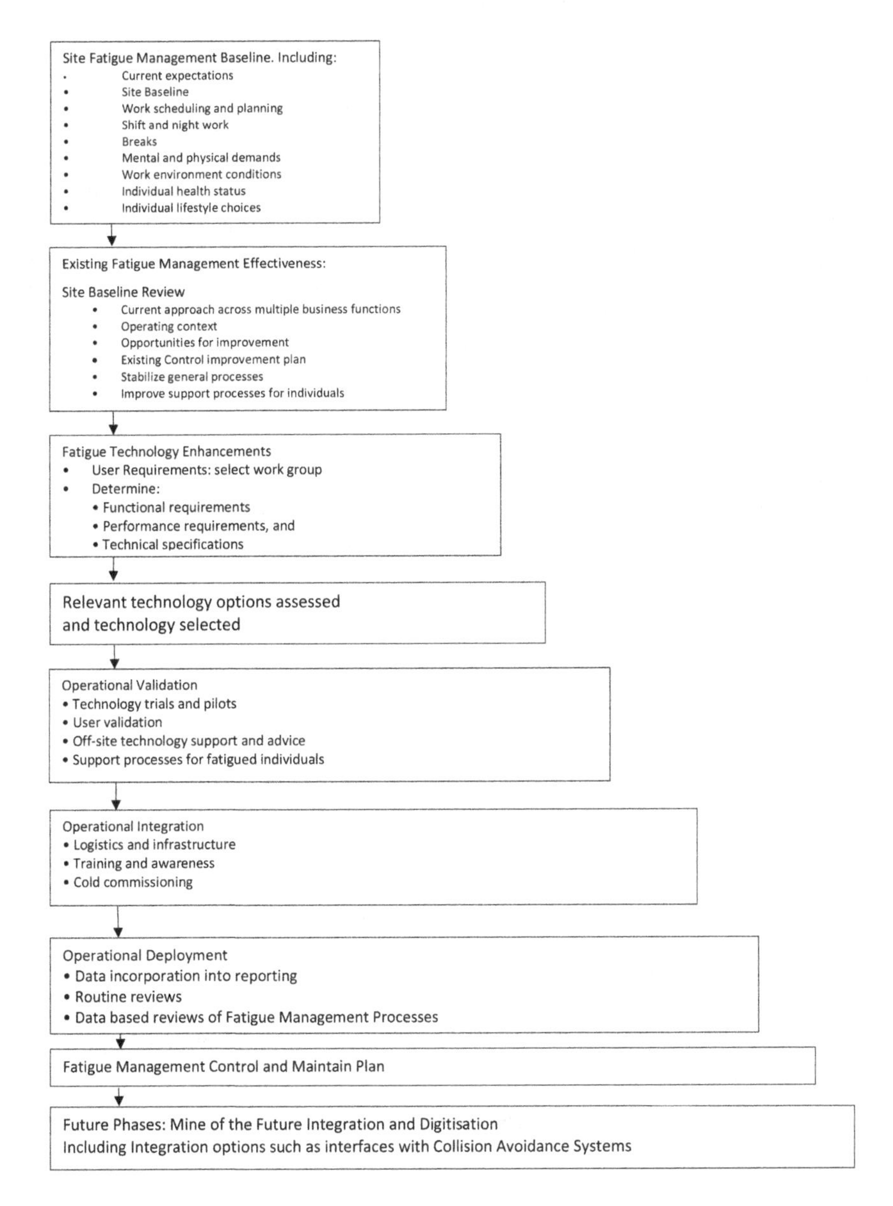

FIGURE 5.1 Glencore Site Fatigue Management Improvement project: work breakdown structure

5.2.4 Outcomes

5.2.4.1 Reduction in Fatigue-Related Incidents

Overall, using the OAS as described above, Glencore have found a year-on-year reduction in fatigue – both in terms of the number of fatigue incidents and their severity. Perhaps the most important finding is ***that only one fatigue-related incident has occurred across all Glencore Coal Assets sites in Australia since the OAS*** was fully deployed.

5.2.4.2 Event Criticality Reduction

A substantial change was realised in the criticality of events being produced between the silent alarm period and the full OAS operation in subsequent years, as shown in Table 5.2.

Overall, the above table shows a downward trend in the criticality of events from the baseline to 2020. For example, whereas 43% of the events were judged as long in the baseline period, by 2020 this was reduced to 5%. A similar trend is also evident for the "moderate" events. Conversely, the percentage of fatigue events classified as "short" increased from the baseline to 2020.

5.2.4.3 Site Comparison

Sites with an approved fitness for work assessment and health intervention TARP showed a lower percentage of long or moderate events compared to those sites without approved fitness for work assessments or Health TARPs. Table 5.3 presents the percentages for different site types.

TABLE 5.2
Event Criticality Reduction

Event Criticality	Baseline	2019	2020
Long	43%	8%	5%
Moderate	16%	10%	7%
Short	39%	78%	86%
Unsafe Distraction	2%	4%	2%

TABLE 5.3
Percentages of long and moderate events by sites with and without approved fitness for work assessment and health intervention TARPs

Event Type	No FFW or Health TARP	Approved FFW and Health TARP
Critical/long	9%	2%
Moderate	10%	6%

Table 5.3 demonstrates the positive impact of fitness for work assessments and health intervention TARPs in terms of a lower percentage of long or moderate fatigue events compared to sites without them.

5.2.4.4 Baseline Data Trend

In terms of fatigue baseline data (without system warnings), the majority of events were produced by a small portion of personnel. For example, at site "C", 43% of all events recorded by the system during one site's baseline period were produced by one operator, as seen in Table 5.4.

Through the identification and management of "at risk" personnel, events by 2020 were distributed more evenly across the workforce and were less critical in nature.

5.2.4.5 Other Findings

In terms of recouping the costs of the system against benefits obtained, as there has only been one fatigue-related incident since the OAS introduction, the system has paid for itself very quickly.

Regarding situations where the system does not work so effectively, it should be noted that it sometimes struggles to identify eye closures through some dark safety glasses and certain prescription glass lens glasses.

Finally, in terms of time-of-day effects for fatigue events, 90% of eye closure incidents occur on night shift and 90% of those occur between 2 am to 6 am. This peak in the early morning corresponds to the typical circadian low-point an operator would experience.

5.2.5 Key Learnings

Four groups of learnings are presented below: human factors, system, fatigue risks, and maintenance management processes.

TABLE 5.4
Baseline data

Site	% of events	Operator Count
Site C	43%	1
Site C	84%	5
Site D	23%	1
Site D	48%	5
Site E	29%	1
Site E	57%	5
Site F	29%	1
Site F	61%	5

5.2.5.1 Project Key Learnings – Human Factors

- Implementation of the system was a six-month human factors journey – from resistance to reliance to response. Potentially counter-productive cultural changes were evident when the system was first introduced, which took a bit of time and encouragement to overcome, which has been aided by the TARPs. In more detail, the cultural journey included:
  1. Resistance:
     a. To the in-cab hardware – camera, IR sensor
     b. Related to video footage privacy concerns
  2. Reliance:
     a. The idea that the monitoring system means that a driver can "Come to work for a rest"
     b. The tendency to "lean on the system" or "drive to the alarm", i.e., over-reliance on the system to the extent that some operators trusted the technology to keep them awake (which is not the purpose of the system)
  3. Response
     a. Intervention through Health TARP (including discussions with mining leadership team, and physical assessment)
     b. health investigations
- Video footage privacy must be maintained; however, operator access can be empowering (e.g., witnessing oneself taking microsleeps at the wheel can be a powerful motivator to seek help or to proactively report one's fitness to work status in the future). There was generally a positive response to the long-term interventions
- Providing system knowledge, including what the system does not do, is very important
- Component failure management – it is imperative that the operator knows when the system is not functional, and to ensure a robust maintenance strategy is in place
- Engage early with key end-users and stakeholder groups – WHSC, SSHR, HSEC, Mining, Maintenance

5.2.5.2 Project Key Learnings – System

- The silent alarm period provided an environment for system commissioning (e.g., avoiding alarm fatigue during initial set-up and calibration of the system)
- Aside from eye closure detection, the system has been invaluable for incident investigation
- The system's capability to detect distraction events (based on head angle) was not utilised
- In-pit wireless communications coverage and bandwidth must be effective
- Data management must be considered (OAS data storage for 2 years)

5.2.5.3 Project Key Learnings – Fatigue Risks

- The silent alarm period allowed for baseline data collection
- Technology is the "last line of defence" and should not be relied on. Better to focus on the Level 1–6 EMESRT controls
- The majority of eye closure events originate from a small percentage of the workforce
- On-shift intervention is not sufficient to change behaviour

5.2.5.4 Project Key Learnings – Maintenance Management Processes

Criticality/FMEA

- Understand the system's failure modes
- Identify component criticality
- Develop maintenance strategies, including repair response times
- Understand what, if any, additional controls are required when the system is critically impaired

Maintenance Process

- Ensure staff, including maintenance personnel, are trained and competent
- Monitor the health of the system in real-time
- Utilise the system's automated sensor failure notification
- Inform the operator of system impairment
- Co-ordinate repairs through your site's defect management process
- Follow up on long-term component failures

5.2.6 Conclusions: Case Study 1

The work presented in this case study illustrated that technology implementation projects are essentially people projects that engender human behaviour changes through the use of technology. Closing the loop regarding operator state and focusing on changing the input (operator behaviour) is an imperative part of the process and ultimately the project's success.

5.2.6.1 Project Benefits

The work described in this case study has produced many benefits:

1. Created an environment where fatigue and inattention events for mobile equipment operators are detected in real-time
2. Provided a layered safety solution to ensure the operator is managed:
    a. at the critical point of impairment through in cab alarm activation
    b. through third-party intervention via positive communications
    c. by conducting an on-shift fitness assessment
    d. in the long term when repeated events occur in an individual

3. Developed a safer mining environment through the reduction and near elimination of fatigue-related incidents for heavy equipment in open cut environments
4. Realised a shift in culture from one where the operator would "power through" impairment (pre-system), to relying on the technology ("drive to the alarm"), to improving proactive reporting of their fitness for work status
5. Developed a procedural avenue for individuals that have multiple reported fatigue events and provide post-shift assistance through Health Intervention Plans
6. Provided capturing resources for incident investigations (with in-cab and forward-facing cameras)

5.2.6.2 Next Steps

As part of an ongoing safety journey, at the time of writing (2022/3) the next things to examine were distraction events and better cab ergonomics. A new study will investigate what drivers are looking at inside the haul truck – especially below the dashboard.

5.2.6.3 Industry Leading and Best Practice Human-Centred Design

As well as being industry-leading, the work contained in this case study was scientifically best practice in terms of following ISO 9241:201 (2019) for a human-centred process. This included carefully exploring the fatigue issue at site, regular worker/stakeholder consultation, evidence-based system selection, employing a staged adoption process, developing effective Trigger Action Response Plans (TARPs), conducting ongoing safety evaluations, and undertaking post-implementation management to further control residual fatigue risks (e.g., through fitness for duty testing). The overall project from a human element perspective is shown in Figure 5.2.

As with all good human-centred design projects, there was a constant focus on users and their work tasks throughout the work described in this case study, including the iterative deployment process and integrating the human and technical aspects of the system.

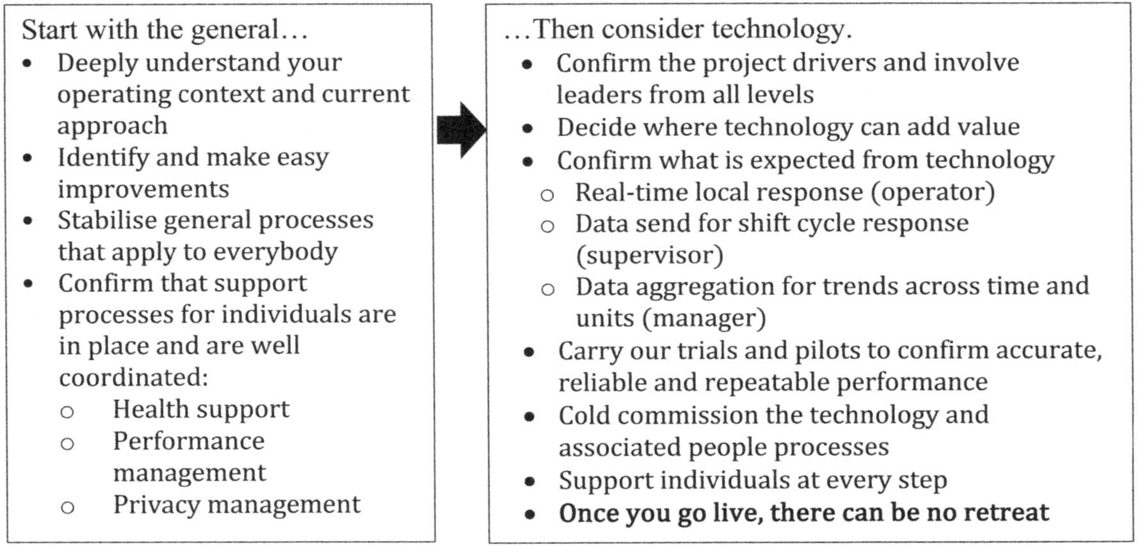

Start with the general…

- Deeply understand your operating context and current approach
- Identify and make easy improvements
- Stabilise general processes that apply to everybody
- Confirm that support processes for individuals are in place and are well coordinated:
  - Health support
  - Performance management
  - Privacy management

…Then consider technology.

- Confirm the project drivers and involve leaders from all levels
- Decide where technology can add value
- Confirm what is expected from technology
  - Real-time local response (operator)
  - Data send for shift cycle response (supervisor)
  - Data aggregation for trends across time and units (manager)
- Carry our trials and pilots to confirm accurate, reliable and repeatable performance
- Cold commission the technology and associated people processes
- Support individuals at every step
- **Once you go live, there can be no retreat**

FIGURE 5.2 Fatigue Management Technology project sequence from a human element perspective

5.3 CASE STUDY 2 – BULLDOZER AUTOMATION: MAXIMISING SAFETY AND PRODUCTIVITY

5.3.1 Overview

This case study focuses on a common type of mobile mining equipment: bulldozers. Following an introduction to the use of bulldozers in mining, and previous human-centred design work with them, the potential to semi-automate bulldozers is introduced. Then, this case study presents two strands of bulldozer automation work. First, experimental studies by McAree and colleagues exploring both perceptual requirements for bulldozer teleoperation and the productivity potential of semi-automated bulldozers: in these studies, the importance of good quality visual information for operator teleoperation was established and it was shown that the semi-automated bulldozers could be as productive as manually operated bulldozers. Second, semi-autonomous bulldozer operations at a mine site in Queensland, Australia, are outlined.

Throughout the material presented in this case study, considering the human element is critically important, for example, when determining the information a remote operator needs to effectively control a semi-automated bulldozer. The case of introducing a semi-autonomous bulldozer at Australian mine sites illustrates many of the deployment and integration challenges typically associated with the introduction of new technology. These included undertaking extensive front-end analysis and assimilating the semi-automated bulldozer into the broader operations at the mine site with a specific focus on considering the human element during the deployment process.

A set of key learnings from this case study are presented. These include how information can best be presented to operators of semi-autonomous dozers, and how a structured approach to mine site deployment of automation that focuses on people and process is vital for successful integration of the technology.

5.3.2 Background

Bulldozers (often simply known as dozers) are commonly used mobile equipment at mines. They are used for ripping/cutting into terrain and pushing material with their blades. Most dozers at mine sites are tracked, are highly powered, and can operate in tight turning circles (Dudley, 2014). Figure 5.3 shows an example of a manual dozer at a mine site.

5.3.2.1 Previous Human-Centred Design Work with Dozers

Due to the ubiquitous nature of dozers at most mine sites, it is not surprising that there has been previous work to improve the working environment and vehicle ergonomics for dozer operators. For example, Horberry et al. (2016) undertook work to improve operator access and egress to the dozer cabin. They noted that a comparatively large number of injuries occur during these activities, partly due to the inadequate design of access and egress facilities on the equipment. In their work, the "OMAT" tool

FIGURE 5.3 Dozer operating at a mine site

developed for mobile mining equipment by the authors and colleagues was used: it combines the application of participatory ergonomics, task analysis, qualitative risk management, and safe design.

Horberry et al. (2016) applied OMAT to this dozer access/egress issue for a global mining company. OMAT workshops related to this specific issue occurred in Australia, Indonesia, and Peru with participants from advisory groups, local management, human factors and safety specialists, and local operators and maintainers. The tasks of getting into and out of a dozer cabin were viewed by the workshop participants at the mine site and also video recorded for additional analysis. An experienced operator and an inexperienced person accessing and egressing the equipment were viewed performing these tasks at the different mine sites. Thereafter, back in the workshop setting, task flowcharts were created for the access and egress tasks, which broke down each task into task steps. Then, risks for each task step were explored and redesign solutions were developed by the group.

One strength of OMAT was giving a site-based mechanical engineer/maintainer a clear understanding of the problem to be solved. The video of tasks being performed was of particular assistance for the workshop group to develop a solution. The redesigned solutions went through several iterations (which initially involved the addition of one handpost, and subsequently two with a refined design), involving user trials in the mine site maintenance areas and further OMAT workshops to help refine them.

The re-design helped an operator to maintain three points of contact. It also helped to constrain their method of access/egress to the designed/prescribed solution (rather than swinging around the handposts). Finally, a subsequent workshop revealed that the handposts should be extended further forwards and backwards to provide additional assistance to the operator.

Overall, the work by Horberry et al. (2016) shows that focusing on end-users and their tasks by means of a structured human-centred process can help produce safer redesigns of mining equipment such as dozers.

5.3.2.2 Dozer Automation and Teleoperation

Due to the versatility of dozers, they are often used in hazardous mining environments. Therefore, approaches to remove the operator from the dozer cabin by means of automation are becoming more popular for safety reasons.

The safety gains of removing a person from the dozer itself must, however, be weighed against productivity concerns: if an automated dozer is less productive than a manually operated one, then uptake of such technology may be hampered (Dudley, 2014). It is therefore important that any automated dozer is capable of operating at a level of performance similar to that of a manually operated one.

Automated/teleoperated systems for dozers are now commercially available. A broad distinction is to separate them into line-of-site vs non-line-of-site systems. In line-of-site, the machine needs to be in direct view of the operator, who controls the machine by means of a hand-held/shoulder-mounted device. By contrast, non-line-of-site systems do not require the operator to be in direct view of the dozer; instead, information is provided to the operator via cameras and other means of giving information such as audio or motion feedback (Dudley, 2014).

Other dozer automation work has characterised a "progression" of control technology. For example, Wolff mining (n.d.) has stages of dozer automation, ranging from line-of-site in which one operator controls one machine through to "semi-autonomous" non-line-of-site operations where one operator controls multiple dozers. After that, the final stage is full dozer autonomy.

5.3.3 Semi-Autonomous Dozers: Studies of Perceptual Requirements and Productivity

5.3.3.1 Perceptual Requirements to Support Effective Teleoperation of a Bulldozer

Work by Dudley et al. (2013) and Dudley (2014) focused on perceptual requirements to support effective teleoperation of a bulldozer. They noted that remote control/teleoperation can provide a means of enabling dozer operators to perform work without being directly exposed to the hazardous environments often experienced during dozer work at mine sites. However, physically removing the operator from the dozer requires the operator to use reduced sensory information to control the machine compared with what would be directly experienced when manually operating the dozer. Dudley et al. (2013) and Dudley (2014) explored the perceptual requirements for bulldozer teleoperation; specifically, what factors are critical to achieving high levels of teleoperation performance and user acceptance.

To undertake this work, the Dudley et al. team developed an enhanced perception cell (similar to a vehicle simulator) capable of high fidelity replication of motion, visual, and auditory cues integrated with an existing bulldozer teleoperation system. The cell allowed systematic analyses of the influence of different individual sensory feedback cues on performance and user acceptance. A series of experiments was conducted to understand better operator perceptual requirements. In particular, they sought to understand what feedback attributes were critical to maximise bulldozer teleoperation performance and user acceptance.

The key result of the work by Dudley et al. (2013) and Dudley (2014) was that visual quality was found to be the dominant factor influencing operator performance. In particular, the provision of depth perception provided the strongest influence on performance for non-line-of-sight dozer teleoperation. Task visualisation to support accuracy and planning was also important. In addition, motion feedback was found to improve aspects of dozer operator performance, but it had no additional benefit in task completion time beyond that provided by enhanced visual quality.

In summary, Dudley (2014) proposed the following ranking of perceptual requirements for enhancing dozer teleoperation performance:

1. Vision
2. Goal-oriented task visualisation
3. Motion
4. Audio

Overall, the team hoped that the knowledge generated through this work might support the maturation of bulldozer teleoperation systems leading to wider uptake and utilisation in hazardous circumstances. The possibility to remove dozer operators from the risks of operating in hazardous environments, whilst still allowing them to work at their full potential, was the ultimate objective of the work (Dudley, 2014).

5.3.3.2 Automated Bulk Dozer Push – Reducing the Cost of Overburden Removal

Subsequent work by members of the same McAree-directed team further examined automated bulk dozer push to potentially reduce the cost of overburden removal (McAree et al., 2017). The collaborative research was undertaken by the University of Queensland and Caterpillar to extend the Caterpillar "Command for Dozing" system that was capable of pivot-push dozing (a bulk dozing method for strip mining overburden removal). The Caterpillar Command for Dozing is a semi-autonomous tractor system for D11T dozers that can perform bulk dozing.

Initially, work by McAree et al. (2017) explored how best to perform the sequence of cuts, pushes, and dumps performed for bulk move pivot-push dozing so that the material moved by the dozer is moved as efficiently as possible. From this, a simulation framework was developed and validated using experimental data collected for a manual push operation. The framework allows the issue of which pivot-push method makes best use of the effective dozer time to be explored.

Thereafter, the work explored how well the semi-autonomous dozer technology performs using the evaluation metric of Effective Time Productivity (ETP; i.e., the cumulative volume of overburden moved to locations beyond the pivot point as a function of time the bulldozer spends in effective operation). The team compared the measured ETP achieved by the semi-autonomous technology at the time of its initial deployment and again after six months, following further technical development.

The initial productivity trial was undertaken when the dozer operators were still in the process of learning how to use the system: it showed the semi-autonomous technology achieved 85% of the expected ETP (as predicted by the framework). Reasons for the gap between expected and actual operation included operator familiarity with the technology as well as technology improvements required to address identified issues.

The second productivity trial conducted six months later, after improvements to the system, showed the semi-autonomous technology has an ETP similar to that for manual dozer operation. Therefore, the work demonstrated that semi-autonomous pivot-push dozing is both technically feasible and its productivity can match manual operation.

The team concluded that the significance of the work to the coal industry is in supporting the realisation of technology for semi-autonomous pivot-push operations and also in providing methods for evaluating the effectiveness of those operations (McAree et al., 2017). Showing that semi-autonomous dozer operation can be as productive as manual operation may be a significant driver in the uptake of such technology in mining (McAree, 2022).

5.3.4 Semi-autonomous Dozer Operations by Thiess

5.3.4.1 Introduction: Thiess' Automation Journey

Thiess is the world's largest mining service provider. This section outlines the ongoing automation journey by Thiess at Lake Vermont and other Australian sites (adapted from Thiess, 2023). Key autonomy achievements with the Thiess dozers include zero autonomy-related injuries. It should be noted that Thiess undertook a large amount of front-end planning prior to dozer deployment, often involving the integration of human element issues. Regarding human systems integration during this autonomy deployment, Thiess note (via WesTrac, 2023):

> In every deployment, we've followed a structured process that involved early engagement with the site and in-depth discussions with the key stakeholders.

Thiess worked with one OEM (Caterpillar) through their journey. As WesTrac/Caterpillar (2023) state:

> As with any technology deployment, the focus is people, process then technology. Investing the time upfront to get people engaged, making sure the messaging is right and clearly communicating the intentions of an automation rollout and what value people can expect from it, will be the keys to a successful project.

5.3.4.2 Discussions with Thiess

The UQ project team had discussions with the SSE at Lake Vermont, and the Group Manager for Autonomous Services at Thiess in April 2023. The following summary represents the state of the operation at that time.

The semi-autonomous dozer installation at Lake Vermont is focused on production-slot dozing (as opposed to typical remote-control dozer use-cases that focus for safety reasons on stockpile operation or other high-risk situations). Thiess sees the opportunity to expand autonomous dozers for rehabilitation applications. But this is more complicated because then the slots would not necessarily be parallel.

Six CAT semi-autonomous tractor systems (SATS) are currently in operation at Lake Vermont – a global partner site for CAT. The base SATS module allows the dozers to track back and forth in a slot, moving earth, with an excavator at one end of the slot. Moving between slots requires remote control. It is a high cognitive load task to decide how to move dozers around, which is where the STAMP software comes in – it relieves the workload by planning how the dozers should be used to achieve the desired goal.

Variability exists in the number of dozers that operators can supervise – top operators can supervise up to five dozers. They suggested that there have been improvements made to the interfaces in response to feedback from operators.

Current training operates on a mentoring model in which a novice sits beside and works with an experienced operator. Operators do not require the same skills as a manual dozer operator, as they are more of a supervisor, but they do need to understand full dozer operation.

Dozers are protected from running into excavators operating nearby by a collision prevention system. This can be overridden when operating in remote control and when dozers have contacted other equipment while in remote mode.

Safety benefits of autonomous dozer and drill operation were noted, as was the opportunity to improve the diversity of work options for those with temporary (or permanent) physical limitations.

5.3.5 Discussion and Conclusions: Case Study 2

The case study presented here describes two different aspects of dozer automation. The first part presented a series of experimental studies exploring perceptual requirements for dozer teleoperation and productivity potential for semi-autonomous dozers. The second part examined the deployment and operation of semi-autonomous dozers in Australia.

Key human element learnings from this case study include:

1. The many benefits that can potentially be realised by automating dozers, including safety (removal of people), productivity, and sustainability outcomes.
2. That dozer automation can take place at several levels, from line-of-site remote control, through to full automation. The level of automation can be tailored to individual mine site requirements.
3. That operator perceptual requirements to support effective teleoperation of a bulldozer can be specified. In particular, Dudley el al. (2013, 2014) emphasised the importance of providing good visual information for effective teleoperation of a dozer.
4. As a driver for further dozer automation work, McAree et al. (2017) showed that semi-autonomous dozer operation can be as productive as manual operation.
5. Discussions with Thiess highlighted their objective of reaching zero injuries and improving workforce diversity (for both, automation plays a vital part), the need to better consider operator training during automated dozer deployment, and using a phased and iterative deployment process that

focuses on people and that captures and applies learnings (e.g., for operator interface design).

5.4 CASE STUDY 3 – LONGWALL AUTOMATION AT GLENCORE AUSTRALIA: TOWARDS SAFER AND MORE EFFICIENT UNDERGROUND COAL MINING

5.4.1 Overview

This case study focuses on work to automate equipment operation in longwall coal mining. Specifically, it presents ongoing work by Glencore Australia at a coal mine in central Queensland.

By means of two site visits (before and after the control room was updated) and off-site interviews with Glencore automation personnel, this case study shows how the company is tackling human element design, deployment, and integration issues by assimilating longwall automation hardware and software from different suppliers. It also presents some of the successes and areas needing more work and outlines some of the next steps to be taken.

From a human factors perspective, the current automation uses remote control and the design of the interface mimics many of the cues used in manual longwall operation. Some of the key human element issues emerging from the automation journey presented here are:

1. **Potential loss of skills** needed for key aspects of longwall operation
2. **Training** for long-term operation of the automated system
3. **Interface design** in the control room (bunker) – particularly better visualisation, more efficient control actions, the use of auditory cues, and interactions between the control room operators
4. **Communications** between the control room and underground – especially developing a clear radio protocol that is suitable for the noisy underground environment
5. **Work organisation** (e.g., work breaks in the control room, shift handovers, minimising distractions in the control room)
6. **Ongoing operator-centred design improvements**: a need for a continuing process to capture operator feedback and regular reviews of the system.

The case study ends by outlining key factors for successful deployment of the automation by Glencore. It is anticipated that these might be useful for other mines planning similar initiatives.

5.4.2 Longwall Automation at Glencore Coal Assets Australia

The information presented in this section is based on interviews with Glencore automation specialists working at the Oaky Creek North mine in parallel with two site visits by two of the authors, described later.

5.4.2.1 History of Longwall Mining at Oaky Creek North

Glencore Australia's mission is for widespread safe and productive remote longwall coal mining. This includes the Oaky Creek North mine. The primary motivation is around human safety – to remove the person from directly operating the longwall underground. They note that additional benefits include the consistency and repeatability of automation in this environment.

Traditionally, longwall mining at Oaky Creek North employed two shearer operators tethered to the shearer via Radio Remotes. The system progressed to one shearer operator with one remote and, with the introduction of remote mining, Radio Remotes were removed altogether and the shearer operator is monitoring the shearer production via "The Bunker" on the mine surface. At the time of writing, the automation of the longwall operation was being completed, including decisions regarding control room layout, communications, training, and operator skills.

5.4.2.2 Longwall Automation Implementation Process

The work at Oaky Creek North is bringing together technology, people, training, and systems. Rather than deploying a single bespoke system, the site automation team is liaising with various suppliers to develop interoperable components to create a system that meets their needs.

Overall, the system is best classified as semi-automation, as a human is still operating the shearer by means of teleoperation. The system still requires "floor steering" where the longwall is mapped by a human underground who reports back the floor corrections to the operator in the control room, who manually enters them into the system. Currently this process is needed to ensure coal is cut within the seam as much as possible. This system eventually may change as technology improves.

5.4.2.3 Human Factors Issues

The following human factors issues were noted from the interviews with the mine-site automation specialists.

Potential Loss of Skills. One key issue noted was potential loss of skills required for control room operator competency. The mine needs a way to validate the competency of the person mapping the floor and to ensure the control room operator can take manual control of the shearer should incidents occur involving conditions which do not permit automation or remote mining (e.g., belt jamming). The current management of skill maintenance is achieved by alternating underground cutting days with working in the control room for both shearer and shield operations. Shield operation is largely the domain of maintenance personnel, particularly mechanical fitters, while shearer operation is undertaken by shearer operators.

Training. The training of new control room operators and maintenance of operational skills remains a challenge for the site. The process to become a control room operator is:

- Have current all appropriate underground tickets. Operators highlighted the need to understand "how work is done" underground to be able to easily use software in the control room.
- A training package is delivered in-house – usually over 2–3 shifts, which includes sitting with an experienced operator following initial training. A supervisor is in the room and can oversee/talk new operators through any situations where they need assistance.

The training package is assessed by the trainer/supervisor and the operator is passed if they are deemed "competent with the system". A key problem noted was the training system is heavily reliant on experienced operators becoming control room operators, but there did not appear to be future planning for training operators who did not have underground experience. At some stage, this pre-requisite will need to change.

Feedback. It is essential to involve those who use the system in the initial design and in modifications as automation progresses. A challenge faced by the site is how best to gather control room operator feedback, including documenting the feedback, and how best to take on board the feedback in future decision-making. Currently it does not appear that control room operator feedback is being captured in a formal manner; however, whilst on site the researchers observed operators readily offered opinions and feedback. Examples of feedback included requests for opportunities to make changes to sequencing, requests for additional cameras for visualisation of "snake areas", and improved radio communication.

Communications. Radio communications were noted to be an ongoing issue. The site was working to develop a more effective method of communication. Communication management is difficult due to underground Wi-Fi issues. The noisy underground environment also creates challenges with conventional communication systems.

5.4.3 First Site Visit, Prior to Control Room Updating

Site visits by two members of the team were conducted in parallel with the above interviews. The purpose of the initial visit was to gain familiarisation with the current longwall automation work, particularly to see the "old" control room and to interview operators. The first visit was conducted in December 2022.

5.4.3.1 Longwall Operations at Oaky Creek North

There are two operators in the control room. Operators reported that initially there was just one screen, but after additional systems were implemented, more screens were needed for monitoring.

The layout has two pods side by side: with two operators (i) shearer operator, (ii) shield operator (who most often is a fitter and turner by trade). Figure 5.4 shows the workstation layout for the two pods that were being used in December 2022.

FIGURE 5.4 Workstation layout at the Oaky Creek North control room in December 2022

Shearer Operations: the following issues were identified with the shearer operator task and workstation layout in the control room before it was updated.

- ***Communications with Underground***. Communication with operators underground (via the DAC system) sometimes is not clear, with operators having to call several times before reaching the underground members of the crew. Additionally, operators had to speak very loudly to be heard by underground operators due to the noisy underground environment. It is understood headphones are going to be trialed underground; ideally, incorporating some form of noise cancelling ability would be extremely beneficial.
- ***Visual Information***. Video footage has good resolution but regularly the cameras have failed connections. Screens do not have continuity of visual access to what is happening underground. The camera locations need review and adjusting. They are frequently dirty, which results in poor image quality.
- ***Ergonomics***. The chairs provided do not provide adequate ergonomic support to the operator. The operators' table was slightly too high for tasks such as writing and typing.
- ***Tasks in the Control Room: Work Scheduling.*** Operators indicated they were pleased they were able to rotate between working in the control room and working underground. When underground ("downstairs"), duties include walking up and down the face, doing corrections, building cogs (timber supports for cut-throughs), and hosing down the face. From interviewing operators, this was considered a break from the boredom of working in the control room.
- ***Tasks in the Control Room: Mental Workload***. Operators indicated the system was relatively easy to learn to use but commented that, although they could take regular breaks, they generally went home more tired than

if working underground because they have been looking at a screen all day. (Anecdotally, one operator reported they no longer watched TV at home after working in the control room, whereas in the past they normally would as a form of relaxation.) Operators also reported the same stress levels/similar workload, as they dealt with things the same way as when working underground, e.g., if a breakdown occurred they would direct maintenance to fix it.

- ***Task Feedback***. Operators indicated they were getting enough information from the screens, but also stated that "it is different to being underground" – they missed the direct feedback from being underground – the vibration of the machine on the floor, and the noise as cutting and when the drums were on the floor. Some felt they could not bring the fundamentals of being underground into the control room – the feedback was missing.
- ***System Over-rides***. Operators are able to over-ride the autonomous system and operate in manual mode; they can skip a "state" (step in the programmed process, which allows them to operate to flow corrections).
- ***Most Challenging Task***. Operators indicated that driving the shearer through gates when the snakes are tight was the most challenging task. While they had good camera perspective, it was not as good as when actually driving from underground.

Figure 5.5 is a photo of the workstation in use.

Shield Operations: Operators expressed concerns with the shield operation task and workstation layout in the control room before being updated.

FIGURE 5.5 Shearer operator workstation

- ***Software***. The CAT system was used (V shield) but the Joy system (LIS) can also be used; however, operators generally considered the CAT system more visually user-friendly. Operators felt the programme has improved over the last 2 years, and that it is still "evolving". Most operators reported the system as easy to learn and use.
- ***Visual information***. The shield operator reported that he would like to see more cameras because, being a fitter/turner, he is trained to be constantly looking at pressures and flow rates. Both shearer and shield operators reported that the cameras get dirty and the images and recordings are not clear.
- ***Diagnostics***. Control room operators reported that the current programme did not provide all the information they would like to have. However, they indicated it did assist with diagnosis/assessment of problems more so than when working underground as they could see what was happening/see the problem (e.g., flow switches/pressure switches). They further commented that, because they could not directly see the equipment from the control room, they were relying heavily on experience and knowledge to keep the system running.
- ***Work arrangements***. Operators reported lack of visual and haptic feedback as problematic. All preferred to alternate control room shifts with underground shifts. Positive feedback regarding control room shifts included a clean and airconditioned work environment away from the dust and noise, and the ability generally to take breaks when needed. Negative feedback was mostly focused on boredom associated with the job. Most reported an increase in tiredness/fatigue at the end of the day.

Figure 5.6 is a photo of the shield workstation in operation in December 2022.

It should be noted that, while the shield operator had access to the ExScan program, the shearer operator did not. The ExScan program detects where humans are underground (and also blockages and shields that have been left behind). It is envisaged this will be added to the shearer control room operator set-up when the new control room becomes operational.

5.4.3.2 Overall Summary of Problem Areas during Initial Site Visit (Prior to Control Room Changes)

As with the interviews with the mine site automation specialists, the key human-related areas identified as requiring further work were:

- *Training*. The operators generally thought there is deskilling (loss of the manual skills from operating a non-automated system) in the control room; however, they thought it was not so difficult to manage the system since they have the experience working underground. The concern is what will happen with new workers, who either have limited or no experience working underground, and how will they learn without practical experience.

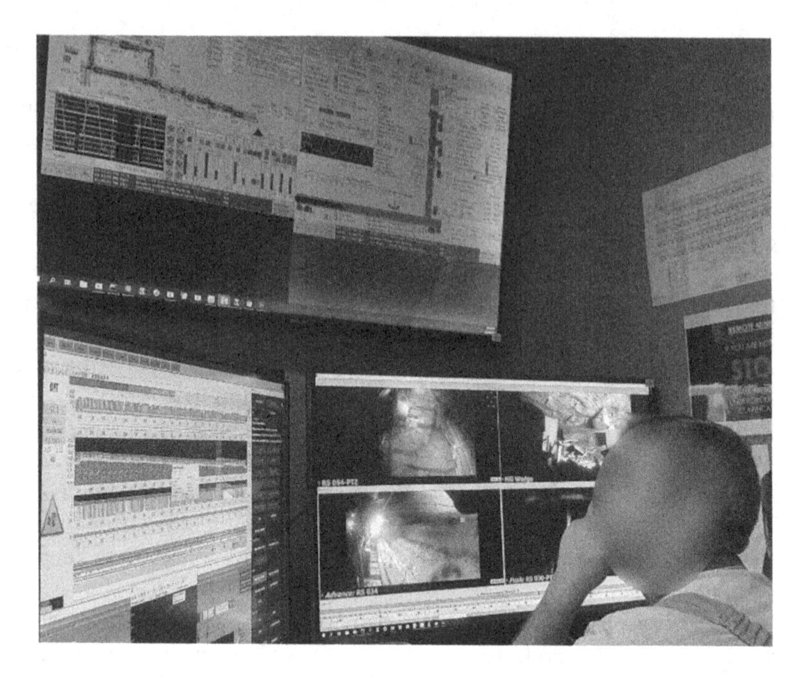

FIGURE 5.6 Shield operator workstation

- *Rostering, shift breaks, and work organisation.* This mostly focused on minimising distractions in the control room.
- *Communications.* This area provides a significant challenge within the current system and it would seem some kind of protocol that standardised messages would be helpful. The communication system needs to be able to accommodate the noisy underground work environment.
- *Control room layout.* At the time of the first site visit, a new control room was at the planning stage. This is the time to decide on the most appropriate equipment, including the best layout of screens and interface design, ergonomic seating, and sit-stand desks. The layout needs to allow for effective interaction between the two control room operators and provide for clear communications with personnel underground.

5.4.4 Second Site Visit, after Control Room Had Been Updated

A second visit by two members of the team took place on 21 June 2023, following the deployment of the new control room. The research team observed the operations and tasks in the new control room and undertook brief interviews with the control room operators.

As previously, the layout has two pods side by side: two operators (i) shearer operator, (ii) shield operator (who is most often a fitter and turner). Figures 5.7, 5.8, and 5.9 below show the revised workstation layout for the two pods in the new control room.

FIGURE 5.7 Revised workstation layout at the Oaky Creek North control room

FIGURE 5.8 Revised shearer workstation layout at the Oaky Creek North control room (left pod)

The new bunker has restricted Swipe access, which reduces control room operator distractions, and adds security for unintended operation of surface control room from unintended personnel. Furnishings have been upgraded to include sit-stand desks; AFRDI rating 6 ergonomic chairs; dedicated air conditioner controls; individual operator light control, front set and rear set lights with independently dimmable control mechanism; and upgraded amenities including fridge and coffee machine. The new control room/bunker is designated as a clean environment, whereas the previous bunker allowed entry to personnel in dirty mine clothing and boots.

The following is a summary of the key points noted.

FIGURE 5.9 Revised shield workstation layout at the Oaky Creek North control room (right pod)

Control Room Layout

1. Overall, the research team considered the new control room was well laid out, with control screens for the shearer and the shields set up on two sides of the room (with shearer controls to the left and shield to the right). Both systems are automated and are equipped to allow supervision and override capability.

Each control room operator monitors seven screens; three being camera views of the underground systems and four screens providing controls and status of the shearer, the ground elevation, and status of the shields.

Work Organisation

2. Many in the crew are now trained to work in the control room and frequently rotate between control room and underground. This helps to address the sedentary work concern that was expressed by operators previously, and from research undertaken in similar environments (e.g., iron ore miners in Sweden). Whilst rotation can reduce the boredom associated with sedentary control room work, it must be recognised that under the current system of control room operator recruitment, the control room operators are firstly operators and are used to manual work, and it may take some time for them to adjust to a new working environment.

Another issue that was noted was the recognition that the control room operators were not necessarily highly skilled in IT and therefore IT specialists/experienced automation supervisors were needed when a more complex automation problem is encountered. In the new control room someone with these skills is rostered on the shift, and located in an adjacent room.

Additionally it must be noted that not all of the crew want to work in the control room. This was described to the research team as older workers feeling more comfortable working underground. This is somewhat consistent with research into age-related acceptance of automation undertaken at Australian coal mines and at an iron ore mine in Sweden.

Interface Design Issues

3. The graphics appear to follow the principle of pictorial realism and the principle of the moving parts. The graphics are colour-coded and appear to support error reduction due to ease of recognition and ease of accessing correct controls. Feedback from control room operators included "the new screens and room set up is much better than the old room in that it is bigger", and "it has a number of wall mounted screens with common information and cameras – this is very helpful". As a possible improvement, one shearer operator commented "we would like a touchscreen and an '*E stop*' button on the desk" (this was also noted in the earlier "SWOT" analysis). According to the operators, they did not consider there was an information overload problem with the displays.
4. Control room operators expressed concern that they manage several software packages for the shearer, for the shield, for the cameras, etc. A beneficial goal may be to try to merge some of these functions into one software system. Similarly, both Komatsu and Caterpillar shields are in use, and control room operators indicated they would like to move to one system, or at least have a better-integrated mix of both.

Safety Improvements

5. The number one stated reason for implementing automation is safety of the workers. Before the system was automated, the shearer operators who worked underground operated the shearer with a wired remote control box, which exposed them to coal dust and excessive noise. Automating the system to the current level allows the two operators to do what they used to do from the control room; however, there are still six to seven crew working at the working face. The underground crew adjusts the panels, changes picks, maintains the equipment, supervises the conveyor, and addresses any blockages that occur.

Productivity Improvements

6. Improvements in productivity have been anecdotally stated; however, the researchers were not able to view the supporting data. Improvement in production was described as related to the consistent and reliable straight-line path that the shearer takes as opposed to the more "irregular" path when operated manually. The straight-line path yields a consistently shorter pass (time). Researchers assume this has been quantified in production statistics. Additional benefits of automating the longwall noted by the control room staff included reduced

maintenance costs due to consistency in equipment operation, and the opportunity to fix issues from the control room with less downtime.

7. Every two to four passes of the shearer, some of the picks on the cutting drum must be replaced (depending on rock versus coal engagement, as the more rock, the more frequently that picks need to be changed). This is currently a heavy manual task with a change-out taking approximately 10–15 minutes. Potentially, technology could be explored that ultimately may allow automated pick changing which would reduce maintainer exposure to injury, and reduce shearer down time.

Future Automation Goals and Ongoing Challenges

8. Overall, the biggest challenge currently experienced in the control room is communication between the control room and the underground crew. It is understood that intended improvements connecting to Wi-Fi underground will help.
9. The ultimate goal of the mine is to become fully autonomous and the mine is progressing towards this outcome. Intermediate goals include:
    - Automate the main gate – which would remove another one or two persons from the hazardous underground operation area. This would assist with setting of the height/creep management and provide consistency, which is key to the process of integrating shearer and shield operation.
    - Remotely map the coal face from the surface, rather than having a person underground walking the face, correcting co-ordinates, and relaying the corrections verbally to the shearer operator in the control room.

5.4.5 Conclusions: Case Study 3

This longwall case study describes an automation journey at one mine site. The journey started with one screen, and incremental additions have been made to the system. The overall goal is to have a system in which all components are well-integrated: in other words, to have effective human systems integration. To achieve this, some of the key human element issues still needing to be addressed are:

- Skills retention needed to operate key aspects of the longwall operation
- Effective training for long-term operation of the automated system
- Refining the interface design in the control room. This includes better visualisation, more efficient control actions (possibly including an "E Stop" button), the use of auditory cues, and improving interactions between the control room operators.
- Communications between the control room and underground – especially developing a clear radio protocol that is suitable for the noisy underground environment
- Work organisation (e.g., work breaks in the control room, shift handovers, minimising distractions in the control room)

- Ongoing operator centred design improvements. There needs to be a continuing process to capture operator feedback and regular reviews of the system.

Despite these ongoing challenges, the overall process resulted in a successful deployment of the longwall automation. Some of the key success factors here were:

- An incremental approach to automation (culminating in the updated control room)
- Involving the operators and other mine site stakeholders in all stages of the process
- Considering the "softer" aspects of automation (such as communications, training, skills retention, and work organisation) in tandem with the "harder" technology aspects
- Working with OEMs and major suppliers (including CSIRO) to integrate, and potentially future-proof various technologies and systems
- Considering both productivity and safety/health concerns as key drivers for the automation deployment
- Regularly reviewing the success of the process (as seen in the mine's post-implementation review)

The potential safety and production benefits from longwall automation are huge. This case study has shown how the automation journey undertaken at this mine can provide valuable lessons for other mines planning to embark on a similar process.

5.4.6 Acknowledgements

The authors would like to thank Glencore Coal Assets Australia and Oaky Creek North site for access to their longwall operations. Thanks to Brad Lucke for organising site access, and in particular thanks to Michael Condie, Automation Projects at Oaky Creek North for invaluable assistance on site and with case study review. The team would also like to thank Project Manager Daniel Becker, Duane Witkowski, and the control room operators and underground crew who generously gave of their time and experience.

5.5 CASE STUDY 4 – COAL MINE SITES IN QUEENSLAND THAT HAVE RECENTLY INTRODUCED AUTONOMOUS HAUL TRUCKS

5.5.1 Overview

Autonomous haul trucks were recently introduced at two coal sites in Queensland: Goonyella Riverside mine and a second mine. Between late 2020 and mid 2022, the authors visited the Goonyella Riverside mine multiple times and the second mine twice. During these visits they made observations of the control room and pit, reviewed documents, and interviewed operators, trainers, managers, and controllers. This case study summarises the findings, starting with the oldest visits undertaken as familiarisation through to the more recent visits which made recommendations for further system integration improvements.

5.5.2 Initial Visit: December 2020

The first visit to Goonyella Riverside provided useful familiarisation to see the Komatsu-built automated system in use on site. It also allowed the team to undertake preliminary observations of the system from the point of view of the control room operators and to see the system in operation in the pit. Through these observations, several focus areas for subsequent visits were identified, including the mental workload of control room operators, the health impacts of control room work, interface design (in both the control room and non-autonomous field vehicles), and the potential for over-trust in the autonomous systems.

5.5.3 2021 Visits

In March and December 2021, members of the team conducted several site visits, encompassing ATH operations at both Goonyella Riverside Mine and a second mine. Overall, the team considered that the technology at both sites was impressive, as was the attention paid to ensuring system safety. Likewise, appropriate technical training and supervision were being provided to those working with the AHT systems. Nevertheless, the team identified three key areas where potential refinements and improvements were proposed: (1) controller workload and health, (2) situation awareness, and (3) interface design.

5.5.3.1 Workload/Health effects

It was observed that control room roles potentially involve high cognitive workload. However, it appeared that controller workload was not solely determined by the number of automated trucks operating, but also the extent of communication required with field roles. This was potentially exacerbated by the requirement for controllers to monitor and respond to multiple communication channels (radio, telephone, in-person). It was noted that the workload fluctuated somewhat unpredictably. Given this, it was recommended that the workload and physical and psychosocial wellbeing of control room staff be monitored regularly throughout the implementation of AHT at current and future sites. The possiblity of using monitoring data to develop a means of predicting controller workload was also raised as potential avenue to improve the effectiveness of workload management (e.g., by first investigating the associations between controllers' scores on measures of cognitive workload and task characteristics, such as time of shift, number of trucks, and environmental factors). Likewise, a similar focus on the workload and wellbeing of field staff was suggested.

In addition, an opportunity for ergonomic assessment of the control rooms was noted. Concurrent monitoring of the musculoskeletal health of control room staff was also recommended.

5.5.3.2 Distributed Situation Awareness

A key requirement for optimal human performance in any system is the maintenance of situation awareness – that is, an accurate and continuous moment-to-moment understanding of the current state of the system and the surrounding environment and, importantly, a projection of likely future states and events that allows timely and appropriate actions to be taken. The AHT systems involve multiple field roles as well as control room roles. The AHT technology itself and all of the people who interact with the autonomous components form a "joint cognitive system". Optimal performance requires optimal decision-making by this team. Timely and appropriate decision-making, in turn, requires the joint cognitive system to maintain ongoing situation awareness. No one person in the system has access to all the information required to do this and so the system may be considered to exhibit "distributed situation awareness". Other examples of distributed situation awareness are found in process control, medical teams, emergency response, aviation, and defense. Maintaining this distributed situation awareness is a dynamic and collaborative process that requies moment-to-moment interaction between team members and the AHT technology, and this poses challenges

Communication between team members is clearly a critical aspect of maintaining accurate situation awareness, as is acquiring and integrating information from the AHT system interfaces. For example, the control room operator does not have access to information about roadway conditions and relies on the field roles to provide that information to allow appropriate actions to be undertaken. Because the controller is required to deal with multiple communication channels (radio, phone, in-person), it is also the case that the field staff may not know if the control room is already attending to another information source when initiating communication. Interpersonal group dynamics are crucial in this situation, particularly rapport between control room and field where interactions are largely virtual, and especially if the controller has limited previous field experience. Accordingly, it was concluded that additional actions to develop team rapport were justified, such as field staff spending time in the control room and vice versa.

Similarly, the design, development, and implementation of non-technical skills training that focuses on teamwork and communication in general, and maintaining distributed situation awareness, in particular, was recommended as one potential avenue to pursue (while by no means a complete solution). For example, it was suggested that such training might fruitfully include facilitated consideration of AHT recordings and concurrent two-way radio logs for particular situations as a means of identifying more effective communication strategies.

The design of the interfaces by which information is exchanged with the AHT technology is also critical to the maintenance of situation awareness. This issue is considered in the next section.

5.5.3.3 Interface Design

The multiple digital interfaces by which field and control room staff exchange information with the AHT technology are critical to supporting both safety and optimal system performance, in large part acting as a means of facilitating the maintenance of situation awareness. While it was indicated that mechanisms exist for providing feedback to the system designers, the consensus was that the response time for any changes was likely to be long and thus the interfaces were unlikely to be able to be altered in the short term. Regardless, it was recommended that a cognitive work analysis of each AHT system be undertaken with a particular focus on the information that is exchanged and the design of the control room and field vehicle interfaces by which that exchange of information is achieved. Where interface limitations are identified, there may be implications for training or procedures that can be enacted as a stopgap in the short to medium term; and the outcomes of the evaluations should be provided as feedback to the manufacturers to inform subsequent system revisions. Such cognitive work analyses may also identify additional training needs.

It was also noted that capturing information gathered during incident investigations associated with the AHT system and analysing the human factors aspects of the circumstances in which the incidents occurred may lead to the identification of potential procedural, system, or environment design improvements; training modifications; or competency assessment requirements to reduce the probability of similar subsequent events.

5.5.4 May 2022 Visit

Three members of the team visited the AHT control room and pit in May 2022. They conducted observations, brief reviews of non-confidential documentation, and non-intrusive interviews with stakeholders. The following summary outlines the key observations and potential suggestions for improvement that arose.

1. **Effective Integration of the Control Room and Pit Operations.** Buliding on the prior recommendations to take actions to develop team rapport, the following specific suggestions were advanced.
   - *Faciliating both control room staff and pit crews to obtain a better 2-way understanding* of one another's constraints. This may be achieved by visits to their opposite work domains, shared team talks, or potentially greater alignment of the control room and pit shifts (so crews get more used to each other).
   - *Bring pit staff into the control room* to see what controllers see and how they work; and take controllers into the pit to see what operations/maintenance staff see and do;
   - *Implement greater standardisation/"technicalisation" of radio communication* (i.e., make it more like Air Traffic Control).
2. **Reducing Error Potential**. It was observed that controller status reports relied heavily on manual input. This increases the controller's workload and potentially increases the potential for transcription errors and missing data. It was indicated to the researchers that this process was a remnant of the manual FrontRunner system and could potentially become an automated function. Hence, better automated controller status reports were

recommended (it was understood that some initial work was being planned here). Similarly, improving delay codes to eliminate the need for extensive typing was recommended as a potential means of improving both controller workload and reporting accuracy.

3. **Investigating System Capacity**. Control room system lags were observed to impede controllers, slowing them down and potentially causing frustration. Since it was unclear whether this was due to the FrontRunner system or computer hardware constraints, an investigation was suggested. It was noted that a potential disconnect between three systems – Modular, FrontRunner, and Tip Head Editor, may also increase controller workload. Finally, it was suggested that the potential for a larger main (zoomable map) screen could also be considered to reduce scrolling/recentring and allow a wider view of activity in the pit.
4. **Enhancing Management of AHTs to Streamline Pit Operations.** It was suggested that, in the medium term, reducing some of the need for LVs in the pit may be possible by changes to procedures and communications to use resources more efficiently. For example, it was noted in the pit that some calls to control could be avoided with changes to the in-vehicle Frontrunner display (thus potentially reducing controller workload and minimising disruptions to haul truck movement). This is because LV operators were often calling control to find out the reason for haul truck stoppages, when the information was already available in the LV Frontrunner display in coded form. It was proposed that one option would be to make this information easier to find and interpret.
5. **Enhanced Mine Planning and Communications.** It was observed that, in the pit, some aspects of mine planning were seemingly still based on full manual operations (e.g., junctions, road widths, turnaround space, and refuelling bay design). It was suggested that mine planning needs to better fit the new capacities and constraints of AHTs (e.g., road design, and moving towards less reliance on the use of taught paths). There was also opportunity to improve communication of site works taking place more than 24 hours in the future).

5.5.5 Goonyella Mine Summary

This section outlines the main findings and potential suggestions for improvement arising from the series of site visits outlined above. These are presented in terms of key themes, arranged into two clusters: (1) ongoing/shorter-term issues, and (2) longer-term challenges that might also form learnings for future automated mine sites.

5.5.5.1 Ongoing/Shorter-Term Issues

1. **Control Room Design Optimisation**. Opportunities to improve control room design were identified. These included investigating system enhancements that would remove log duplication by controllers and improve the quality of screen graphics. One suggested interface improvement was to

enable pop-up status boxes to be be moved around the screen so that they do not block other important information. Another suggestion was to give the controller the capability to type in important announcements that reach relevant operators.

The importance of collaboration and good communication between controllers and pit staff was highlighted as critical for safe production. AHT controllers spent time in the pit as part of their training to understand the perspective of an operator and/or pit supervisor and suggested there was also an opportunity for pit supervisors and/or operators to come into the control room to get a controller's perspective.

2. **Equipment Maintenance/Breakdown**. Unplanned maintenance impacts production. For example, when a truck becomes disabled and requires maintenance, it sometimes blocks a haul road. The controller will need to programme alternate routes into the system for all trucks that must pass the disabled truck. Maintenance must be closely coordinated between the maintenance department and the controller. The rest of the operation must also be updated on the status in each maintenance situation. However, there may be ways of doing it differently (e.g., increase the planned maintenance checks for trucks and have the maintenance checks done in pit, bring in a mobile maintenance workshop with all equipment/crib attached, geofence-off the contained area).
3. **Training**. A suggested training improvement could see preloaded scenarios for the new controllers to practice on – using a simulator. Trainee controllers could part-task in simulator exercises for aptitude testing and familiarisation before they start in the control room.

5.5.5.2 Longer-Term Challenges

4. **Distributed Situation Awareness**. The onsite control rooms had a team supervisor that led the team of controllers and had the overarching "big picture" of the total operations in the pit to make decisions when controllers needed to escalate a concern. However, given the complexity of the system, it is impossible for any one individual to have complete situation awareness at all times. Nevertheless, it was considered feasible to take steps to increase distributed situation awareness by the mine site team.It should be emphasised that controllers have very difficult jobs that require a high level of skill. Not everyone is capable of this job, and there is so much information provided to the controller that there is a risk of information overload. This makes distributed situation awareness a must, and more data therefore should be collected, processed, and delivered to the right people at the right time.
5. **Complacency/Over-trust of the Automated System**. Another key human factor is that the workers may become complacent and over-rely on the automated system. Although there is decreased likelihood of incidents with AHTs as they can "see around the hill/around corners" and can anticipate collisions 300m away, the potential for an incident still exists, especially as there are still many pieces of manually operated equipment on site.

6. **The Role of Rules/Legislation**. Regarding legislation, one view expressed was that new legislation is not helpful at this stage because AHA is new and needs more maturity before risks can be classified an calculated appropriately. Further discussions with State-based regulators regarding regulations, codes, and guidelines are occurring.
7. **The Role of LVs**. A fundamental review of the role of LVs in a mature, automated pit may be needed in the future. Many dozers, graders, shovel-operators, water trucks, and light vehicles are in the pit with the AHTs so it is essential for the communication system to work well. There is a critical need for effective, positive communication with each other when required, especially with the controller. From this, it is necessary to rethink the risk patterns in the pit, particularly how LVs are operated. A view expressed was that the biggest risk remains "human error" (e.g., being run over by HV) while sites cannot completely remove the human element (e.g., operators of graders, dozers, and water carts). One control in place was a separate LV road around the AHT domain to reduce light/heavy vehicle interactions. Also, reducing some of the overall need for LVs may be possible by changes to procedures and communications to use resources more efficiently.

5.5.6 Second Mine Summary

Two members of the team paid a final visit to the second mine control room and pit in March 2023. As with earlier trips, they interviewed AHT supervisors/superintendents and a senior manager, conducted observations, and held brief non-intrusive talks with control room operators. This section outlines the mainfindings and conclusions for that mine.

The control room appeared to be well-organised. Staff workloads did not seem to be excessive at the time of the site visit (the final day of the crew's shift cycle). The following six recommendationswere identified as areas for potential improvement to the systems in operation at that time, and were intended to help improve safety and productivity.

1. **Better Capturing Learnings**. The supervisors/superintendents noted that some learnings from the second AHT deployment process were not as well captured as they could have been. Likewise, learnings from other automated mine sites had potentially been not fully utilised during the second mine AHT deployment. A formalised process across mine sites to better capture learnings could provide additional benefits.
2. **Improved Training and Recruitment**. Potential improvements to training were identified, including: having more trainers available, better provision for workforce upskilling, preloaded upset scenarios for new controllers to practice on, and use of a simulator. It was also recommended that recruitment and aptitude testing should focus on key skills required in the control room, both technical and non-technical (including as communication and teamwork), with increased psychometric testing at the beginning of the recruitment process.

3. **Consideration of a Lead, Overall Controller**. It was recommended that consideration should be given to providing a lead, overall controller in the control room since no single person had the "big picture" of the total operations in the pit and control room. This would allow one person to have a "larger world" understanding (i.e. high-level situation awareness) of what is happening. It was understood that this was not fully the shift supervisor's responsibility at this site.
4. **Bridging the Control Room/Pit Disconnect**. It was observed that, due to operational challenges in the control room and the pit, there may, at times, have been some disconnect between the two. It was suggested that more work could be done to bridge the gap between control room and pit staff to help reduce any potential gaps. It was understood that informal secondments/visits did sometimes take place between the pit and control room, and it was suggested that formalising this for new operators and as refresher training may be of benefit.
5. **Greater Use of Camera Information**. It was understood that a large amount of camera information waspotentially available, but only a small amount was used and displayed on large screens in front of the two controllers. It was suggested that better use of the cameras could occur, with information better displayed and the screens potentially relocated. It was recommended that a review be conducted regarding what camera information would be beneficial for different roles in the control room, with additional screens installed to display this information.
6. **Review the Control Room Layout**. The control room layout had controllers/builders facing one direction, and Maintenance M1, technical, and the shift supervisor facing another. This may not be a suitable layout for some situations (e.g., when a controller needs to liaise with the technical operator or M1, they need to turn around and raise their voice or walk across the room) and so it was recommended that the layout of the control room be reviewed.

5.6 CHAPTER CONCLUSIONS

The four case studies present automation and associated human system integration, across a diverse range of mining operations: these include surface haulage and underground mining. Drawing all this together, Table 5.5 presents 10 key lessons learnt, and at which stage in the mine lifecycle each lesson might be most applicable.

It is anticipated that these 10 key learnings will be valuable for other mines considering new technology and automation. They are useful as pointers to mine sites to better understand and manage the human element in automation. Also, it is stressed that open reviews are important about what aspects of automation did not work as well as planned for the industry to progress towards a more integrated and automated mine of the future.

TABLE 5.5
Key learnings from the case studies

	Key learnings	Examples and further information
At mine planning, technology design, and procurement stages		
1	**The importance of Human Systems Integration (HSI) for mining.** Promote and ideally regulate the consideration of human systems integration as a key component of the acquisition/procurement process.	As seen in the AHT case study, the prior development of a human system integration management plan for mining automation procurement and deployment is crucial. HSI principles and, in particular, the use of human-centred design methods, should be applied to technology procurement and deployment of all levels of complexity and scale. Doing so is crucial to achieving the anticipated benefits of automation.
2	**Data**. Better capture and use of data to understand automation needs, to quantify benefits/issues and for technology selection. This should include maintenance needs.	The importance of the collection and use of appropriate mining data has been seen throughout these case studies. For example, it was used to select the most suitable operator alertness system and supported the business case for the performance benefits of automated dozers. Obtaining a good understanding of technology maintenance needs before it is procured is of key importance. It was seen in the AHT case study, in terms of the light vehicles to support the automated haul trucks: there were regular problems with the availability, design, and crewing of the light vehicles – in part because their important role was not fully understood until after the automation was deployed at the mine site.
3	**End-users.** The need to engage end-users and stakeholders early and regularly.	End-user and other stakeholder engagement was seen in the longwall case study to help improve the design of the automation control room to create a heathier work environment.
Prior to, and at, site deployment		
4	**Workload**. The importance of managing operator workload and organising the work appropriately.	As seen in AHT case study, the workload of operators in the newly commissioned control rooms was often excessive. This was likely to lead to burnout, errors, and staff turnover. Ways to reduce workload include better communication protocols and automated reporting functions that the control room operators often performed manually.
5	**Interface design.** The importance of interface design and effective control room layout.	The dozer and AHT case studies showed the importance of designing user-friendly interfaces and effective control room layouts. Likewise, the longwall automation case study showed how an improved control room layout helped create a healthier and potentially more productive work environment.

(*Continued*)

TABLE 5.5 (CONTINUED)
Key learnings from the case studies

	Key learnings	Examples and further information
6	**Training and skills.** The issues of training, loss of manual skills and recruitment.	The importance of appropriate training, preventing the loss of manual skills, developing the right skills for automation, and ongoing recruitment were evidenced through the case studies where there was a growing recognition that different skills for workers will be needed with automated systems. The AHT control room work showed that long-term planning to recruit and retain suitable workers and to develop appropriate skills for automation is vital for success.
7	**Communications**. The importance of effective communications in control rooms.	The importance of effective communications, especially in control rooms where there is a potential for disconnects to the pit, was seen in AHT case study. Possible approaches to help improve communications include considering the need to have a lead controller in the control room with an overview of the whole operation, and better standardising radio protocols, especially between the control room and pit.
8	**Cultural changes**. Managing cultural changes with automation	As seen in the operator awareness case study, managing cultural changes by deploying automation at a mine is vital. In this case study, it required the operator awareness system users to move from resistance to reliance to response.
9	**Adoption process: Integration at site**. Integrate automation with mine site procedures and systems using a staged and structured deployment process. Longer-term automation goals are considered	The introduction of new technology/automation at a mine site has to be integrated with existing procedures, manual equipment, and systems. Successful integration of new technology into existing mine sites was seen in the operator awareness case study where health TARPS were developed to support the fatigue detection system. Those mines with health TARPS had more effective operator awareness systems than those sites that had not yet introduced them. A staged approach was in evidence in the AHT case study – in which functions to be automated were prioritised: for example, first haul trucks then later automating the main gate.
At the post-deployment stage		
10	**Feedback loops.** The importance of ongoing operator-centred design improvements and feedback.	Developing a structured process to better capture learnings and end-user feedback allows for better redesigns, procurement decisions, and healthier workplaces. This was seen both in the longwall and AHT case studies where the feedback obtained helped to redevelop control rooms that were ergonomically more suitable. Such feedback loops may also be useful for other mine sites to learn lessons from the earlier adopting sites.

6 Risk Assessments for Mining Automation *(by guest author Professor Maureen Hassall, Minerals Industry Safety and Health Centre, Sustainable Minerals Institute, The University of Queensland, Australia)*

6.1 INTRODUCTION

Risk is the uncertainty that matters because it can affect the attainment of objectives that are valued. In mining, risk is often determined by the likelihood and consequence of experiencing adverse events such as those that result in harm to humans, environmental damage and/or production losses. In many mining jurisdictions, mine safety legislation and regulations require mine operators or duty holders to identify and eliminate or manage so far as is reasonably practicable the risks that could cause harm to the health and safety of workers. So, to avoid events that could lead to adverse health and safety outcomes, mine operators use risk management processes to prospectively identify and address potential health and safety risks. The risk management processes used typically follow those outlined in ISO31000 (International Organization for Standardization, 2018) as illustrated in Figure 6.1.

The importance of risk assessment and risk treatment processes has been highlighted in investigation findings from both mining and non-mining autonomous technology incidents. Examples include:

- The NTSB findings from automated Uber trial vehicle crash in 2018 that killed a pedestrian, which included the following: "contributing to the crash were the Uber Advanced Technologies Group's (1) inadequate safety risk assessment procedures, (2) ineffective oversight of vehicle operators, and (3) lack of adequate mechanisms for addressing operators' automation complacency" (National Transport Safety Board, 2019, p. v).
- The outcomes from the ATSB investigation into the 2018 Western Australian autonomous train runaway and derailment included the following statements: "Rail transport operators should therefore ensure that they conduct

DOI: 10.1201/9781003380887-6

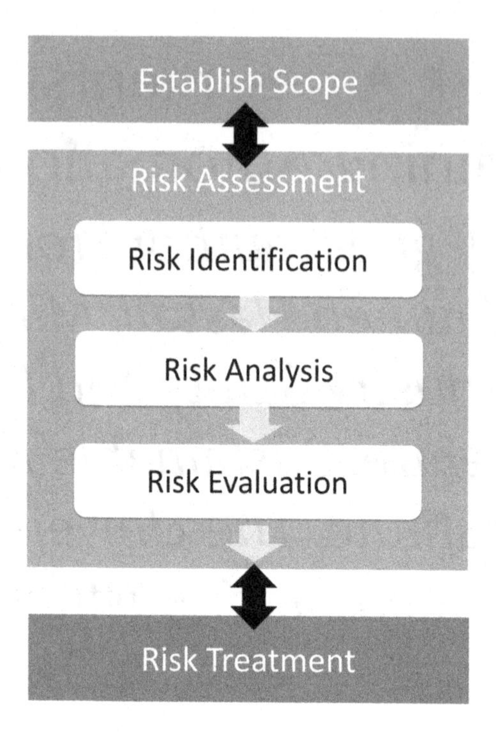

FIGURE 6.1 ISO31000 Risk management process

thorough risk assessments to ensure that relevant causes and hazards associated with runaway events are identified and managed. In addition, rail transport operators considering changes involving the integration of complex systems should utilise a systems engineering approach to identify hazards and then manage risk to ensure that the railway's operations remain safe, so far as is reasonably practicable. Rail transport operators must then ensure the preventative controls mitigating the hazards will be effective in managing the risk. They also need to place adequate emphasis on critical controls to signify their importance and ensure that the rail safety workers who are required to implement procedural controls clearly understand why the specified actions are required" (Australian Transport Safety Bureau, 2022).

- The investigation into the 2015 collision between an autonomous mine haul truck and manned vehicle found that the management of change processes associated with vehicle route planning and assignment were inadequate and therefore recommended that "responsible persons at mine sites using autonomous mobile equipment are reminded of the importance of identifying, monitoring and reviewing hazards associated with the interaction of manned and autonomous mobile equipment" (Department of Mines and Petroleum – Resources Safety, 2015, p. 2).

Technique options that can be used in risk assessments are described in International Standard IEC31010 (International Electrotechnical Commission, 2018). Specifically, this standard provides guidance on the selection and use of different techniques for assessing risks in a wide range of situations. The techniques that are strongly applicable to risk identification are various brainstorming techniques; Failure Modes and Effects Analysis (FMEA); Failure Modes, Effects, and Criticality Analysis (FMECA); Hazard and Operability studies (HAZOP); Hazard Analysis and Critical Control Points (HACCP); Human Reliability analysis; and Structured What-If Technique (SWIFT). These techniques were developed many decades ago when systems were simpler, less interconnected, and more electromechanical in nature so they focused more on technology failures. They were not designed to capture interaction, novel and emergent risks associated with digitalization, new automated and autonomous technologies nor with dysfunctional interactions that can occur between software-hardware and humans within highly connected, complex socio-technical systems where accidents happen due to deviations in performance rather than individual component failures (Dekker, Cilliers, & Hofmeyr, 2011).

New techniques have been developed to help undertake risk assessments on complex, interconnected, digitalized socio-technical systems that are being increasingly introduced into modern mining operations. One such technique is the System Theoretic Process Analysis (STPA) which is normative or descriptive in nature. Another technique is Strategies Analysis for Enhancing Resilience (SAfER) which is formative in nature. These techniques are discussed in more detail next.

6.2 STPA

The STPA technique leverages Nancy Leveson's System-Theoretic Accident Model and Processes (STAMP), or STAMP work (Leveson, 2004, 2012). STAMP is based on the premise that systems failures occur due to the failure to control deviations from safe operations caused by design flaws, component malfunctions, software errors, design flaws, external disturbances, and dysfunctional interactions among system hardware, software, and human components. The STPA utilises the fundamentals of STAMP to identify risks associated with a failure to adequately control or enforce behaviour-related safety requirements of the system (Leveson, 2012; Stringfellow, 2010). STPA uses the operating process part of the STAMP model to identify potential areas where there could be inadequate control in order to improve control design. The process starts with developing a model describing the control structure (Leveson & Thomas, 2018). Simplified versions of these diagrams can be drawn to highlight human-system interactions as shown in Figure 6.2 for transferring fuel from a ship to an onshore tank.

Prompts or guidewords to the components and interactions to identify and assess the deviations from safety operations that could occur in order to determine the controls that need to be implemented and enforced to ensure safe operations. (Leveson, 2012; Leveson & Thomas, 2018). Example guidewords include: "No control command given"; "Inadequate commands are given"; "Commands given too early/late"; and "Control action stops too soon or lasts too long". A worksheet can be developed to capture guideword analysis for each control action as shown in Figure 6.3. This figure shows the analysis done on two of the tank filling control actions – "operator starts filling process" and "operator stops filling process".

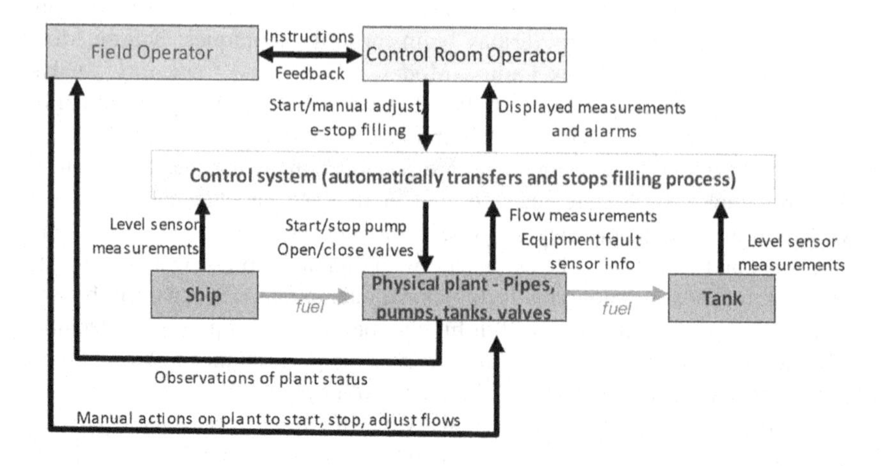

FIGURE 6.2 Human-system interaction diagram for transfer fuel from a ship to a tank onshore

Control action	Control Action NOT GIVEN	INCORRECT Control Action IS GIVEN	Control Action GIVEN AT WRONG TIME TOO SOON/EARLY	Control Action GIVEN AT WRONG TIME TOO LATE	Control Action GIVEN IN WRONG ORDER or FOR WRONG DURATION	Potential Consequence(s) and Significance (High priority - must address, Med priority - should address, Low priority - monitor for change, Negligable - No further action required)	Possible causes of unsafe control action	Assessment and recommendation for improving design, or controls or orgnaisational systems
Operator starts filling process using system "start" button	Operation doesn't execute start fill button on filling station computer	Operator selects stop (not start) filling in computer	Operator executes start command much too early	Operator executes start command much too late	Not applicable	Significant unsafe action = If started too soon, tanker may not be properly parked, connected and earthed which could lead to loss of containment and/or the introduction of an ignition source making fire and explosion possible >> HIGH PRIORITY	Operator distraction or lack of competency, poor shift handover.	Install interlocks of forcing function based checklist that requires a specific operator to check system and tanker is correctly set up prior to pressing start.
Control system stops filling process when tank full or alll fuel transferred	Computer doesn't execute stop fill button on filling station computer	Computre slows, restarts, increases flow (instead of stopping) filling	Computer executes stop command much too early	Computer executes stop command much too late	Not applicable	Significant unsafe action = If not stopped or stopped too late then this could lead to loss of containment due to overfilling tank which could cause fire and explosion >> HIGH PRIORITY	Sensor failure, communication failure, software error, loss of power. fill station valve/pump failure	Install two independent system with separate comms and power that automatically stop if tanker High High Level reached or no liquid following in pipe or mass balance reconciliation shows filling complete or system is loosing liquid

FIGURE 6.3 Example of STPA worksheet

The STPA process tends to be normative or descriptive in nature in that it tends to be based on work as intended (e.g., as described in procedures or imagined by analysts) or work as observed and described to capture work as done when observed or done by those consulted. The SAfER process, described next, is formative because it is based on the range of ways work could be done in both normal and abnormal conditions by a diversity of people.

6.3 SAFER

SAfER is based on the premise that humans can behave and respond within systems in a range of ways depending on their assessment of the situation, the risks, the difficulties, and time pressures (Hassall, 2013; Hassall et al., 2014). Therefore, to ensure system safety, the system design needs to be designed to promote human situation awareness, decisions, and performance that allow humans to successfully anticipate and respond to deviations in system performance and to tolerate and manage other variabilities so that it will not result in adverse outcomes or unsuccessful outcomes (Hassall, 2013).

The SAfER process involves two main steps. The first step is a Situation Assessment analysis that seeks to identify and improve situation awareness. It involves identifying the indicators of safe/unsafe operations and then determining the design improvements that could or should be made to make these indicators very obvious or salient. They need to be very easy to perceive, very easy to comprehend in a correct, valid, and timely manner, and very easy to project or forecast into the future. Situation assessment indicators may be hardware focused, software focused, human focused, and/or worker focused. Examples of situation assessment factors for the automatic tank filling scenario are shown in Figure 6.4.

The second step in the SAfER process involves identifying the range of ways people can perform work in normal and abnormal situations using eight generic strategy prompts and then determining the design improvements that help operators select and enact the preferred strategies while tolerating or preventing other strategies in a manner that prevents adverse outcomes. These eight generic strategies combined with example responses and analysis are shown in Figure 6.4. An important part of this analysis is recognizing that some design measures designed to help humans manage normal operations can hinder the adaptable human responses needed to manage abnormal event situations (Flach et al., 2003; Leveson, 2012; Vicente, 1999). The reverse can also be true. Examples of this can be seen in alarmed systems where alarms that help in normal situations can go into alarm floods that overwhelm and/or confuse operators in abnormal situations. Conversely, alarms that are intended to detect and alert operators of abnormal situations can lead to false alarms which desensitize operators causing them not to respond to true abnormal situations. Other examples exist in the analysis of emergency responses that occurred during the Piper Alpha and Deepwater Horizon events. Further information on SAfER can be found in publications by Hassall (2013, 2022).

Critical Situation Assessment Factors	List the factors that need to be monitored to ensure safe operation	Design interventions for making the critical factors more salient to controllers
Plant/process factors	- Status of pipes, connections, valves, pumps etc involved in fuel transfer - Status of safety-critical equipment required to monitor fuel transfer (e.g. instrumentation and communication systems) - Contents, flowrate and levels within pipes and tank - Indicators of fuel mass is being lost/unexplicitedly gained in system - Current vs projected fill time	- Real-time mass balance (reconcilation of what it coming off ship is going into tank) and projected tank fill time - Instrument functional/fault alarms - Colour coding of different fuels as it flows through through pipes and when it is stored in tanks
People factors	- Real-time location and status of field operators and other personnel doing safety critical work on/near ship, pipelines and tank. - Availability and vigilance of control room operator (and backup) to start and oversee fuel transfer	- GPS tracking of operator displayed on same interface that shows fuel flows - AI enabled camera monitoring of safety critical parts of process, including control room operator, that detects and alerts anomalies - Live communication system between control room and field operators (e.g. radio system)
Context factors	- Current and forecasted adverse weather conditions - Presence of other vessels in/around wharf - Other work/activity in/around ship, pipelines and tank	- Lightning warning system - Non-routine work tracking system

FIGURE 6.4 Situation assessment part of SAfER analysis

6.4 ASSESSING RISKS IN MINING SYSTEMS WITH AUTOMATION

Research was conducted in 2021 to determine what combination of risk assessment techniques delivers the most effective means of identifying risks associated with human-system interactions in remote and autonomous mining operations (Hassall et al., 2022). This research involved workshopping with mining industry professionals the application of Preliminary Hazard Analysis (HAZID), Failure Mode and Effects Criticality Analysis (FMECA), Strategies Analysis for Enhancing Resilience (SAfER), and System Theoretic Process Analysis (STPA) techniques on automated systems in mining and collecting perceptions on the usability and usefulness of these techniques in identifying and assessing safety risks associated human-system interactions. The findings from this research were summarized graphically as shown in Figure 6.5, which highlights that SAfER was perceived to be the most effective

Generic Strategy	What decision/actions might be associated with this generic strategy might be used in normal operations to prevent unwanted events and in abnormal operations to mitigated adverse outcomes and why a person might choose this strategy?	Potential Consequence(s) and Significance (High priority - must address, Med priority - should address, Low priority - monitor for change, Negligable - No further action required)	Should design promote, prevent or tolerate strategy and why?	Design improvement recommendations
Avoidance = Not done, defer, or forget to do	For normal operations: - Control room operator doesn't start loading or stops loading because plant or people not assessed as ready	Productivity will be lost but Safety will be maintained Med Priority	Promote - This is strategy will help ensure safety operations of system	1. Ensure operator has E-stop that she/he can and will manually activate from control room
	For abnormal operations: - Control room operator does not address LOC because it isn't detected by instruments	Spill could become large harming environment and/or form into vapour cloud that explodes - High Priority	Tolerate because cannot prevent but need to ensure use of strategy does lead to catastrophic outcome	2. Vapour/liquid monitoring systems to detect leaks 3. Mass balance, expected vs actual displays to highlight loss of containment issues
Intuitive = automatic response, done without explicitiy or deliberately using thought processes	For normal operations: - Control room operator start loading assuming everything is ok (e.g. plant and instrumental is functional, connections from ship to tank are correctly made, fuel type and quality are to spec etc)	Catastrophic LOC event could occur resulting in fire, explosion and/or significant enviromental harm High Priority	Tolerate because cannot prevent but need to ensure use of strategy does lead to catastrophic outcome	3. Manual or automated checks with alarm process that is interlocked with unloading pumps so system cant start until check on safety critical equipment, connections, etc are done
	For abnormal operations: - Control room operator assumes only one LOC event is occurring	Catastrophic LOC event could occur - High Priority	Tolerate because cannot prevent but need to ensure use of strategy does lead to catastrophic outcome	As per 2. above
Arbitrary-choice = guessed, scrambled haphazard or panicked response	For normal operations: - Control room operator guesses which piping and tank to use	Potential mixing of products - Low priority as only product quality issue	Prevent with known industry standard designs	4. Color coding of tanks and pipes to indicate which are ULP, diesel etc and heights shown on display (see pic)
	For abnormal operations: - Control room operator guesses size of spill and how best to respond	Spill could become larger harming environment and/or form into vapour cloud that explodes - High Priority	Prevent with known industry standard designs	5. System has ability to do real time reconciliations and display LOC volumes and automatically shutting down filling system if LOC occurs
Imitation strategies = copy how others do it or copy what has worked in the past	For normal operations: - Control room operator copies how previously unloaded ULP (but could be different tank(s), different volume, etc)	Potential mixing of products and/or spill due to overfill - Product mixing is low priority but overflow is high	Tolerate because cannot prevent but need to ensure use of strategy does lead to catastrophic outcome	As per 4. above with addition of 6. Overfill SIS system to automatically shut off filling if tank is full
	For abnormal operations: - Control room operator copies previously used LOC event response (but this event could involve different fuel, different location, different cause etc)	Overfill spill could become larger harming environment and/or form into vapour cloud that explodes - High Priority	Prevent with known industry standard designs	7. Automated vapour/liquid loss detection systems connected to automatic shutdown, automatic deluge and sitre emergency response system so operator has help addressing abnormal situation
Cue-based strategies = select Chosen Option using the Observed Info/Cues and Predict Consequences results	For normal operations: - Control room operator closely monitors unloading process on screens	Desired outcome - High priority	Promote This is strategy will help ensure safety operations of system	8. Employ forcing function technology to ensure unloading on progresses when operator monitoring screens (e.g. eye tracking, acknowledge buttons etc)
	For abnormal operations: - Control room operator looking for and acts on 'weak signals' of abnormal operations (chronic unease)	Desired outcome - High priority	Promote This is strategy will help ensure safety operations of system	9. Camera and interface systems allow operator to do "deep dive" interrogations
Compliance-based strategies = following procedures as they are written/practiced	For normal operations: - Control room operator follows SOP	If the SOP is correct and known this should be promoted	Promote This is strategy will help ensure safety operations of system	10. Embed SOP within CRO monitoring system as a checklist process so detailed procedural reading not required (integrate with 8.)
	For abnormal operations: - Control room operator follows Emergency response plan	If the plan is correct and can be acted on intuitively, this should be promoted	Promote This is strategy will help ensure safety operations of system	11. Create ERP checklist (similar to aviation) to help operator expediently activate and monitor emergency response
Analytical Reasoning strategies = using analytical thinking to reason out the best way to perform task	For normal operations: - Control room operator goes back to first principles and checks and double checks everything before starting the unload process	Could significantly delay unloading if process takes time. If done correctly risk is low	Tolerate because cannot prevent but need to ensure use of strategy does lead to catastrophic outcome	12. Give operator camera line-of-sight, a smart control system and a field operator to expedite checks without undermining the quality of them
	For abnormal operations: - Control room operator thinks about and develops own emergency response	Delayed response could like worsen emergency situation - High	Prevent with known industry standard designs	13. Conduct regular emergency response drills so reaction to LOC events becomes a well practiced response.

FIGURE 6.5 Examples of generic strategy prompts and analysis done on tank filling

technique and STPA the most efficient to learn even though these techniques were new to the participants. This suggested a hybrid approach. The remainder of this chapter draws upon this research to demonstrate, using a case study example, a hybrid approach for assessing risks in mining systems with automation.

6.5 HUMAN-SYSTEM INTERACTION RISK ASSESSMENT TECHNIQUE

To identify and assess risks associated with human-system interaction, research suggests that the following steps might deliver the most useful outcomes:

1. Set the scope using a scope table to clearly articulate what people, locations, equipment, activities, timeframes, external considerations, and other assumptions are included or excluded from the analysis. Documenting the scope complies with the ISO31000 step of setting the context. It also helps clarify the boundaries and focus areas of the risk assessment in a manner that will inform readers and users of the risk assessment in the future. Additionally it can be referenced in the change management process to ascertain whether the proposed change will have a significant enough impact on the scope of what was included or excluded enough to warrant a re-examination of the risk assessment. The documented scope table can be supplemented with operational diagrams, site layout images, and other drawings and diagrams that help people consider the whole system in sufficient detail to be clear on what is/isn't included, to identify all safety-critical situational awareness factors and human-system interactions, and to identify and assess all the material risks.
2. Develop a human-system interaction/control diagram. The human-system interaction diagram should identify safety-critical personnel/roles and key system components that interact in ways that can influence safety. This typically takes the form of a block diagram where the roles and components are identified in blocks. Annotated arrows are used to highlight the safety-critical interactions and communication flows that should or could occur between the blocks. The annotations of blocks and arrows should clearly describe the safety-critical function of the role, component, communication, or interaction.
3. Undertake a situation assessment (step 1 of SAfER analysis) that involves indicators that need to be perceived, comprehended, and projected in order to detect and assess the safety status of the system. The situation assessment should also determine risks associated with indicators being absent/overlooked, misleading or incorrectly perceived, comprehended, or projected and the design interventions needed to make the high-risk indicators more salient and easier to comprehend and project. Situation assessments should be done for all humans controlling or impacted by the safety of the system. It can also be done for equipment operating within the system.
4. Conducting a strategies deviation analysis (combination of STPA and SAfER step 2) on the safety critical interactions identified in the human-system interaction diagram. This analysis involves identifying deviations in safety critical human-system interactions using prompts from both STPA and the SAfER categories of strategies and identifying the causes and consequences

TABLE 6.1
Scope table for autonomous surface haulage (sourced from Hassall et al 2022a)

Attribute of scope	Included	Excluded
P: People involved in risk management or potential impacted if risks are not managed	Employees, contractors, OEM personnel and visitors with access to automated haulage areas and/or equipment including: - Controllers who manage automated fleet - People who maintain the virtual map of the mine - Controllers who manage other/manual pit operations - Personnel who operate manned vehicles and equipment in pit - Autonomous and manned vehicle field support, service and maintenance personnel - Other personnel that enter into pit (e.g. supervisors, geotechnical, mining and other technical specialists, etc) - IT & communications people - Visitors – authorised and unauthorised	People outside the autonomous areas in the pit and outside the pit control rooms.
L: Locations or areas where the risk exist or that could be impacted if the risk event materialised	Surface lease areas accessible to automated fleet (operating in autonomous/semi-autonomous modes) - Roads - Active mining areas – including loading, hauling and dumping area - Park up areas and zones where trucks transition from autonomous to manned - Refueling areas	Off-active lease areas, exploration, care and maintenance sites Onsite new equipment delivery and commissioning areas
E: Equipment and plant (e.g. tools, vehicles, fixed processing plant, infrastructure etc)	Autonomous haul trucks Load units (manually driven) Water carts (manually driven or autonomous) Road maintenance equipment (manually driven) Load and dump area cleanup plant and equipment Vehicles that fuel and service in-pit equipment BOMB/MMU/MPU truck Other ancillary equipment (e.g. lighting towers, communications and network hardware etc) Light vehicle fleet	Aerial vehicles (manned and unmanned). Other stationary in-pit equipment eg sumps/ pumps, crushers and conveyors. Process plant area

(Continued)

TABLE 6.1 (CONTINUED)
Scope table for autonomous surface haulage (sourced from Hassall et al 2022a)

Attribute of scope	Included	Excluded
A: Activities (e.g. operations, maintenance, startups etc)	Loading, hauling, dumping, in-field troubleshooting, equipment cleaning, roads and other work areas where automated haulage equipment can access. Human-system interactions include: - Manual digging and loading of trucks - Shift and break changes for dig, dozing and cleanup equipment operators - Manual inspecting, mapping and surveying of mine involving people in light vehicles and on foot - Haul road watering involving manned vehicles or autonomous - Haul road maintenance involving manned vehicles - Hauling and dumping at ROM, berms (manual/auto), paddocks, over-edge both in autonomous and manual modes - Cleanup around load, roads and dump areas performed by manually driven equipment - In-pit manual intervention to inspect, troubleshoot and/or reset autonomous trucks includes people on foot approaching trucks - Queuing of vehicles at loading and dumping areas - Refueling of vehicles – both manual and autonomous - In-field servicing of vehicles – both manual and autonomous - Parkup of vehicle – both manned and autonomous at active mining areas, on haul roads, at cribs, workshops etc - Accessing autonomous equipment – in-field - Transitioning autonomous equipment from manned to automated mode and vice versa - Control room oversight of autonomous fleet Mine control of entire operations including authorising entry, dispatching, allocating water truck and ancillary fleet duties etc.	Delivery, unloading and commissioning of new vehicles. Decommissioning and removal/disposal of old/ written-off vehicles. Areas outside autonomous zones e.g. Drill and blast areas. Rehabilitation activities
T: Timeframe (e.g. time based exposure info, timezone info and how far into the future)	Continuous operation – 24 hours per day, 7 days a week, all seasons of the year in Australian climatic conditions. Consideration should include shift and break changeover processes Adverse climate conditions – wet weather, lightning, dust, fog, ice, and extreme heat/high temperatures.	

(Continued)

TABLE 6.1 (CONTINUED)
Scope table for autonomous surface haulage (sourced from Hassall et al 2022a)

Attribute of scope	Included	Excluded
S: Known risk scenarios that need to considered	Potentially fatality or severe injury resulting from unsafe human-automation interaction associated with: 1. Manual and autonomous driven vehicles operating in same area includes manned light, ancillary and heavy fleet operating where autonomous haul truck are operating includes • Ensuring digital maps accurate reflect the actual/live/current status of the physical operational area • Operators setting accurate assignments from field – bays, spot points etc for autonomous vehicles • People understand where automated zones exist and different zones within automated area • People understanding what the autonomous vehicle status and intention 2. People parking light vehicles and being on foot in areas where autonomous trucks are operating includes: • People performing normal mine duties e.g. installing signage, operator change-outs, vehicle maintenance & refueling • People approaching and moving away from autonomous haul trucks that has stopped/ broken down in the field. • People approaching and moving away for vehicles transitioning trucks between manned and autonomous mode in designated area 3. Human responding/not responding to control system guidance, safety alerts and exceptions • Operators remote controlling from the field • Control room controllers overriding control system or misinterpreting/incorrectly handling exceptions • Control room controllers setting assignments from control room – bays, spot points etc for autonomous vehicles	Autonomous vehicle fire. Autonomous vehicle – autonomous vehicle collision Manned vehicles incidents not involving autonomous vehicles Interactions with automated water trucks – future scenario Interactions with automated dozers – future scenarios

(*Continued*)

TABLE 6.1 (CONTINUED)
Scope table for autonomous surface haulage (sourced from Hassall et al 2022a)

Attribute of scope	Included	Excluded
E: External considerations including known weather and climate related conditions (e.g. heatwave, tsunami, bush fire, earthquake), fauna or flora issues (e.g. dangerous wildlife) and human/societal considerations (e.g. cyber, political, social)	All weather conditions for operations. Autonomous system has to deal with wildlife encroachment from land and air. Cyber security and physical security of autonomous operations. The supply of utilities and communications to and around site.	Weather conditions not experienced by operations. Supply of non safety- critical componentry and service

of these deviations to determine their risk ranking. Then determining the design improvements needed to address high-risk ranked items in a manner that will help ensure systems safety. In determining deviations, causes, consequences, risks, and design improvement options, the analysts should research relevant literature, incident investigation information, and consult with operations personnel, subject-matter experts, and other relevant stakeholders to ensure a well-informed and comprehensive analysis is performed.

5. Synthesising the identified design improvements derived from steps 3 and 4 into a table of prioritised actions expressed in specific, measurable, achievable and assigned, relevant, and timebound (S.M.A.R.T) manner.

To demonstrate this process, a case study analysis for autonomous haulage operations in a surface mine is presented. This also builds on and extends the autonomous haulage case study presented in Chapter 5. The step 1 scope of autonomous haulage operations is shown in Table 6.1. This table provides an example of how considerations that were included and excluded from the risk assessment can be clearly articulated in a tabulated form that is easily referenced and can easily be checked for completeness and accuracy and in management of change processes.

The step 2 human-system interaction diagram is shown in Figure 6.6. This diagram only focuses on control of processes around autonomous trucks being loaded. It highlights the safety-critical people, equipment, communication flows, and interactions and their functions associated with manoeuvring autonomous trucks into the loading position under the manned loading unit (e.g., shovel, dragline, loader, etc.), loading the trucks and releasing the loaded trucks to autonomously transport loaded

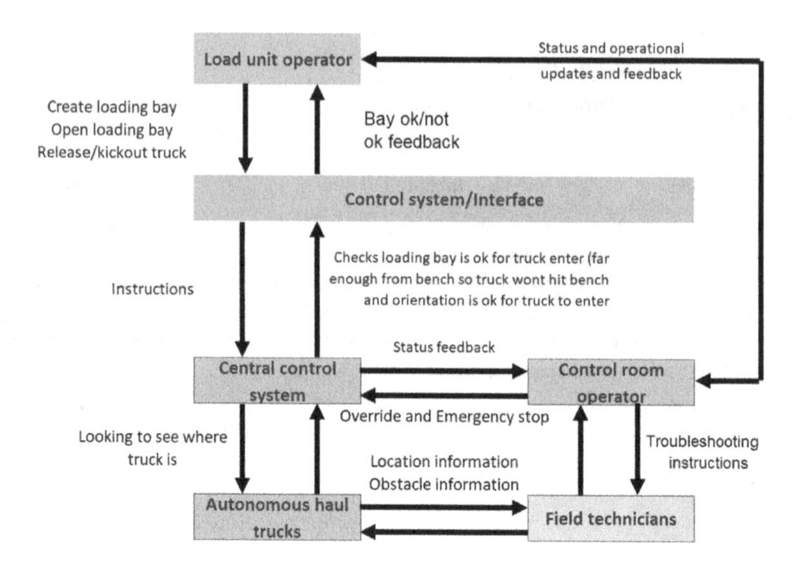

FIGURE 6.6 Human-system interaction diagram associated with loading of autonomous trucks

material to dump location. The diagram can be adapted to include other activities that might be relevant at a given site such as dozer or loader clean-up activities that can occur around a large dig unit. The diagram can also be extended, or additional diagrams be created that capture functions associated with autonomous haulage between load unit and dump site, dumping, refuelling operations, etc.

The step 3 situation assessment analysis performed for the load unit operator only and drawn from Figure 6.6 human-system interaction diagram is shown in Table 6.2. This analysis highlights the key indicators that the load unit operator needs to perceive, comprehend, and project; the causes, consequences, and risk rank associated with any deviations in situation assessment indicators; and the design interventions that would make the more safety critical indicators easier to perceive, comprehend, and project.

An example of the step 4 strategies deviation analysis conducted on the load unit operator safety critical interaction of creating and opening a loading bay so an autonomous truck can park next to the loader and be loaded is shown in Table 6.3. This analysis highlights that there are a range of ways that the operator might perform this task and reasons and consequences associated with the different strategy options. The analysis also highlights that some of the actions should be promoted, others should be tolerated and others should be presented and design improvement options for doing so.

TABLE 6.2
Situation assessment analysis for load unit operator.

Situation Assessment Indicators	List the indicators that need to be monitored to check for safe/unsafe operation?	Causes of indicator being absent/overlooked, misleading or incorrectly perceived, comprehended or projected	Consequences of indicator being absent/ overlooked, misleading or incorrectly perceived, comprehended or projected	Risk Rank	What design improvements could make these indicates easy to perceive, comprehend and project into the future?
Plant/ process factors	Loader Status = Presence/ Absence of critical alarms - e.g. alarms which indicate status of brakes, lifting related hydraulic systems, vehicle stability, and functionality of communication systems to autonomous and through radio system to control room	No/broken/ malfunctioning alarm systems. Poor alarm design leading to false, flooding, confusing, non-salient, non-intuitive alarms.	Worse case = loader instability leads to roll-over or communication issues cause truck to collide with loader. Both scenarios are potential fatality scenarios	High	** Future research is required to identify appropriate human-centred, ecologically designed interfaces and leading design requirements for alarm systems and to investigate independent backup means of determining loader status
	Mode/status of individual haul trucks and overall autonomous haulage system is in = autonomous, faulted, manual	No/failed/incorrect/ confusing indication on truck and on autonomous interface in loader.	Worse case = Truck does not respond to commands entered into autonomous interface and responds in a different way than loader operator wants which could lead to truck entering or leaving load bay when loader and/or bay not ready which could lead to equipment damage	Medium	Salient mode lights on physical trucks and on interface showing status of trucks and system

(*Continued*)

TABLE 6.2 (CONTINUED)
Situation assessment analysis for load unit operator.

Situation Assessment Indicators	List the indicators that need to be monitored to check for safe/unsafe operation?	Causes of indicator being absent/overlooked, misleading or incorrectly perceived, comprehended or projected	Consequences of indicator being absent/ overlooked, misleading or incorrectly perceived, comprehended or projected	Risk Rank	What design improvements could make these indicates easy to perceive, comprehend and project into the future?
People factors	Presence of other fixed/ mobile plant entering, within and leaving loading area Presence of people - authorised and other entering, within and leaving loading area	Reliant on visual checks of area and checking on the interface that can be confounded by visibility issues (e.g. sun glare, blind spots, poor lighting at night). Interface indicators could also be malfunctioning or loss of communication	Worse case = loader operator unaware of presence of person/ equipment and they could be in impact zone of loader and/or truck. This could end in collision resulting in fatality and/or severe damage to assets	High	Proximity/obstacle/intruder detection using AI based vision that predicts potential and detect actual incursions and alarms or stops equipment based on collision potential. This system could be backed by tradition lidar/ radar technologies
	Loader operator status related to mental and physical health levels including fatigue, vigilance, hydration etc	Reliant on self-assessment and/or non-valid or lagging indicators	Worse case = distracted, erratic, unsafe loader operation which could lead to equipment damage or the operator illness	Medium	In cab visual and biometric real-time monitoring fed to operator and if no response alerts control room

(Continued)

TABLE 6.2 (CONTINUED)
Situation assessment analysis for load unit operator.

Situation Assessment Indicators	List the indicators that need to be monitored to check for safe/unsafe operation?	Causes of indicator being absent/overlooked, misleading or incorrectly perceived, comprehended or projected	Consequences of indicator being absent/ overlooked, misleading or incorrectly perceived, comprehended or projected	Risk Rank	What design improvements could make these indicates easy to perceive, comprehend and project into the future?
Context factors	Condition of loading bay e.g. clean, obstructed, vacant, occupied	Reliant on visual checks of area that can be confounded by visibility issues (e.g. sun glare, blind spots, poor lighting at night).	Worse case = could lead to truck entering or leaving load bay when bay not ready which could lead to equipment damage if truck detection system did not pick up obstacle	Low	Proximity/obstacle/intruder detection using AI based vision and tradition lidar/ radar technologies
	Current and forecast weather conditions	In correct prediction of weather due to system inaccuracies, loss of communications, loss of trust	Worse case = lightning strike blowing out truck tires during loading and fragments striking and killing operator	High	Lightning tracker with all site stop processes should be implemented.

TABLE 6.3
Strategies deviation analysis

Strategies deviation analysis	What plausible decision/actions related to this generic strategy could be used in the system being analysed?	Causes of indicator being absent/ overlooked, misleading or incorrectly perceived, comprehended or projected	What consequences might result if people adopt this strategy?	Risk Rank	Should design promote, prevent or tolerate strategy?	What design improvements would improve response strategies during normal, abnormal and unexpected situations?
Avoidance = Not done, defer, or forget to do	Instruction to create or open loading bay not given.	Bay could be unsafe to enter. Truck may not be present. Load unit operator may be distracted/ taking break. Site may be stopped due to weather or emergency	Trucks will not enter loading zone. Trucks will not be loaded and production will be delayed.	Low	Tolerate	1. Have backup system that seeks response from operator if loading delay is unusual.
	Truck has issue manoeuvring into bay but loader operator ignores it.	Operator doesn't check truck as doing other function or cannot see issue or does not comprehend issue	Truck may not position correctly in load bay resulting in spillage or may collide with load unit causing asset damage and injuring operator	High-	Prevent	2. Trucks to be fitted with collision avoidance systems. 3. Trucks to be fitted with fault detection, shutdown systems.

(Coninued)

TABLE 6.3 (CONTINUED)
Strategies deviation analysis

Strategies deviation analysis	What plausible decision/actions related to this generic strategy could be used in the system being analysed?	Causes of indicator being absent/ overlooked, misleading or incorrectly perceived, comprehended or projected	What consequences might result if people adopt this strategy?	Risk Rank	Should design promote, prevent or tolerate strategy?	What design improvements would improve response strategies during normal, abnormal and unexpected situations?
Intuitive = automatic response, done without explicitly or deliberately using thought processes	Loading bay is opened when loading bay is free of obstacles	This would be in compliance with procedures	Truck will be able to travel into loading bay without being obstructed.	High+	Promote	4a. Provide obstacle/ intruder detection cameras to monitor loading bay areas and use AI analysis on vision to support operator decisions about status of loading bay area (e.g. show green when free of obstacles)
	Loading bay is opened when area is occupied by rocks and/or other vehicles	Operator doesn't check or doesn't see due to visibility issues	Entering truck may collide with rocks or other vehicles which could be manned, leading to fatality	High-	Prevent	4b. Provide obstacle/ intruder detection cameras to monitor loading bay areas and use AI analysis on vision that prevents the bays being opened if there are obstacles present. Same as 2. above

(Coninued)

TABLE 6.3 (CONTINUED)
Strategies deviation analysis

Strategies deviation analysis	What plausible decision/actions related to this generic strategy could be used in the system being analysed?	Causes of indicator being absent/ overlooked, misleading or incorrectly perceived, comprehended or projected	What consequences might result if people adopt this strategy?	Risk Rank	Should design promote, prevent or tolerate strategy?	What design improvements would improve response strategies during normal, abnormal and unexpected situations?
Arbitrary-choice = guessed, scrambled haphazard or panicked response	Loading bay is created in guessed or arbitrary spot	Operator impairment and/ or there was error/issue inputting instructions into interface	Truck may collide with load unit causing damage and potentially injuring/killing loader operator	High-	Prevent	5. System design should prevent loading bays being created in positions that could lead truck to collide with loader.
	Loading bay is created or opened too soon	Distracted, rushing operator	Entering truck may collide with rocks or other vehicles which could be manned, leading to fatality	High-	Prevent	Same as 2. above
	Loading bay is created or opened too late	Operator may be unfamiliar with task and interface may be confusing or hard to use. Operator may be distracted	Trucks will not enter loading zone. Trucks will not be loaded, and production will be delayed.	Low	Tolerate	Same as 1. above

(Coninued)

TABLE 6.3 (CONTINUED)
Strategies deviation analysis

Strategies deviation analysis	What plausible decision/actions related to this generic strategy could be used in the system being analysed?	Causes of indicator being absent/ overlooked, misleading or incorrectly perceived, comprehended or projected	What consequences might result if people adopt this strategy?	Risk Rank	Should design promote, prevent or tolerate strategy?	What design improvements would improve response strategies during normal, abnormal and unexpected situations?
Imitation strategies = copy how others do it or copy what has worked in the past	Loading bay is created and opened without checking status	Loading area has not been obstructed by rocks/obstacles	Entering truck may collide with rocks or other vehicles which could be manned, leading to fatality	High-	Prevent	Same as 3. above
	Loading bay is created or opened before trucks arrive but obstacle (e.g vehicle or fallen rock) enters load bay then truck	Loading area left opened for long duration without monitoring which is done/ encouraged to so as to no delay production	Entering truck may collide with rocks or other vehicles which could be manned, leading to fatality	High-	Prevent	Same as 2. above

(Coninued)

TABLE 6.3 (CONTINUED)
Strategies deviation analysis

Strategies deviation analysis	What plausible decision/actions related to this generic strategy could be used in the system being analysed?	Causes of indicator being absent/ overlooked, misleading or incorrectly perceived, comprehended or projected	What consequences might result if people adopt this strategy?	Risk Rank	Should design promote, prevent or tolerate strategy?	What design improvements would improve response strategies during normal, abnormal and unexpected situations?
Cue-based strategies = select Chosen Option using the Observed Info/Cues and Predict Consequences results	Loader operator checks load area before creating load bay	Trained, vigilant, unimpaired operator who is not rushed.	Loading bay will only be created in area where there are no obstructions.	High+	Promote	Same as 4a. Above
	Loader operator checks condition of truck before opening load bay	Trained, vigilant, unimpaired operator who is not rushed.	Only trucks observed to be fault-free will instructed to enter loading bay.	High+	Promote	6. Loader operator interface should clearly show availability and fault status of trucks. 7. Loader operator should have ability to park trucks with visual or system faults and get them actioned.

(Coninued)

TABLE 6.3 (CONTINUED)
Strategies deviation analysis

Strategies deviation analysis	What plausible decision/actions related to this generic strategy could be used in the system being analysed?	Causes of indicator being absent/ overlooked, misleading or incorrectly perceived, comprehended or projected	What consequences might result if people adopt this strategy?	Risk Rank	Should design promote, prevent or tolerate strategy?	What design improvements would improve response strategies during normal, abnormal and unexpected situations?
Compliance-based strategies = following procedures as they are written/practiced	Loading bay is opened when loading bay is free of obstacles	This would be in compliance with procedures	Truck will be able to travel into loading bay without being obstructed.	High+	Promote	Same as 4a.
	Loading bay is not opened if no fault-free trucks available or if obstacles present.	This would be in compliance with procedures	Trucks will be prevented from entering bay if not safe to do so.	High+	Promote	Same as 7. above 8. Operator has ability to organise bay cleanup
Analytical Reasoning strategies = using analytical thinking to reason out the best way to perform task	Loading operator thinks through all loading bay creation options before proceeding.	Operator may be unfamiliar with task and interface may be confusing or hard to use. Operator may be distracted	Trucks will not enter loading zone. Trucks will not be loaded and production will be delayed.	Low	Tolerate	Same as 1. above

7 Conclusions: Human Systems Integration for Mining Automation

7.1 INTRODUCTION

As we have seen in the preceding chapters, the introduction of mining automation has great potential to reduce safety and health risks by removing people from hazardous situations. However, automation changes the tasks performed by people and may introduce new hazards. Current standards and guidance materials pay insufficient attention to the integration of humans and technology during the implementation of automation in mining. For optimal system functioning, the tasks undertaken by people in the system must be designed taking human abilities and limitations into account. Human-systems integration aims to ensure that human-related issues are adequately considered during system planning, design, development, and evaluation. This chapter describes the elements of a human-systems integration plan addressing each of the six HSI core domains: staffing; personnel; training; human factors engineering; and safety and health.

7.2 HUMAN SYSTEMS INTEGRATION

Human systems integration (HSI) refers to a set of systems engineering processes originally developed by the Defence industry to ensure that human-related issues are adequately considered during system planning, design, development, and evaluation (International Council on Systems Engineering, 2015).

For example, the USA Department of Defence (2022b) requires programme managers to undertake a combination of risk management, engineering, analysis, and human-centred design activities including:

- the development of a human systems integration management plan
- taking a human engineering design approach for operators and maintainers
- task analyses
- analysis of human error
- human modelling and simulation
- usability and other user testing
- risk management throughout the design life-cycle
- developing a training strategy

DOI: 10.1201/9781003380887-7

And obliges lead systems engineers to:

> use a human-centered design approach for system definition, design, development, test, and evaluation to optimize human-system performance … Conduct frequent and iterative end user validation of features and usability … (and) … ensure human systems integration risks are identified and managed throughout the program's life-cycle.
>
> **(USA Department of Defense, 2020)**

The processes described aim to ensure that human considerations are integrated into the system acquisition process. The importance of including human systems integration subject matter experts throughout the acquisition programme is made explicit. It is notable that, in contrast to the existing mining automation guidance (Chapter 4), system safety is considered to be a domain within human systems integration.

Similarly, the USA National Aeronautics and Space Agency (NASA) requires human systems integration to be implemented and documented in a Human Systems Integration Plan. The plan identifies the steps and metrics to be used throughout a project life-cycle, and the methods to be undertake to ensure effective implementation. Effective application of human systems integration is understood to result in improved safety and health, increased user satisfaction and trust, increased ease of use, and reduced training time, all leading to higher productivity and effectiveness.

The methods have progressively diffused to civilian industry. For example, the USA Federal Railroad Administration defines human systems integration as a "systematic, organisation-wide approach to implementing new technologies and modernising existing systems". It combines methods, techniques, and tools designed to emphasise the central role and importance of end-users in organisational processes or technologies. Useful human systems integration guidance has been provided for the acquisition of complex railroad technologies (Melnik et al., 2018).

7.3 HUMAN SYSTEMS INTEGRATION FOR MINING SYSTEM ACQUISITION

Human systems integration incorporates human-centred analysis, design, and evaluation within the broader systems engineering process. That is, human systems integration is a continuous process that should begin during the definition of requirements, continue during system design iterations, and throughout commissioning and operation to verify that performance, safety, and health goals have been achieved.

A framework for human systems integration during implementation of new technology in mining is presented in Figure 7.1. Six domains relevant to mining are defined: staffing; personnel; training; human factors engineering; safety; and health.

7.3.1 STAFFING, PERSONNEL, AND TRAINING

"*Staffing*" concerns decisions regarding the number, and characteristics, of the roles that will be required to operate and maintain the joint human-automation system.

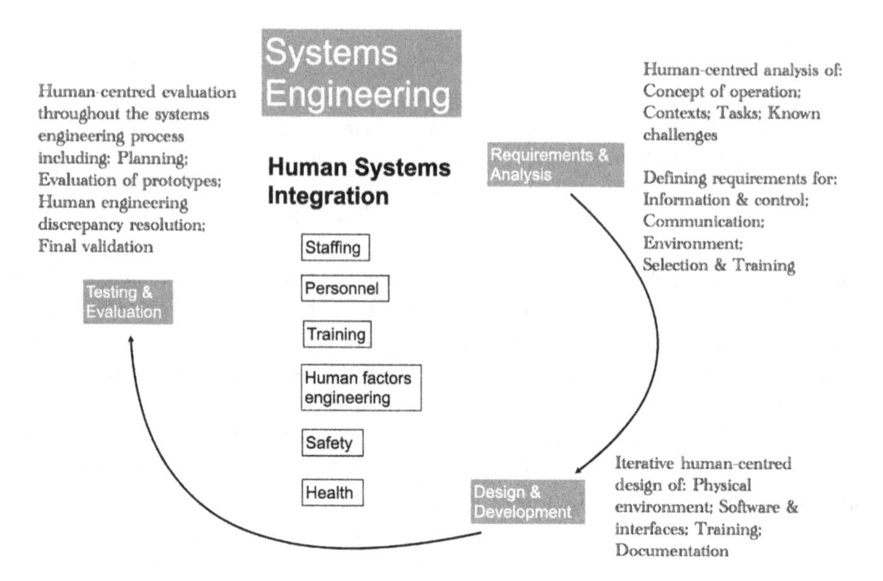

FIGURE 7.1 A framework for human systems integration during implementation of new technology in mining

Decisions here may well require consideration of the outcomes of investigations in other domains particularly where workload issues are involved.

The "*personnel*" and "*training*" domains concern, respectively, the related issues of the characteristics of the personnel who will be selected to fill those roles; and the extent and methods of training, and competency assessment, involved in preparing personnel to obtain and maintain competencies (knowledge, skills, and abilities) required for safe and effective operation and maintenance of the joint human-automation system. Rather than decreased, training requirements for operators interacting with highly autonomous systems are likely to be increased to ensure the operation of the automation is fully understood. For example, automated system controllers need to understand the following: system hazards and logic, and reasons behind safety-critical procedures; potential results of overriding controls; and how to interpret feedback. Skills for solving problems and dealing with unanticipated events are also required. Emergency procedures must be over-learned and frequently practiced.

Instructional system design models (Gordon, 1994) exemplify the application of human factors principles to training. In essence, such models involve front-end analysis steps (analysis of the situation, task, equipment interface, trainees, training needs, and resources, leading to definition of the training functional specifications), followed by design and development steps (training concept generation, training system development and prototyping, and usability testing) and system evaluation steps (determining training evaluation criteria, collection and analysis of these data, and subsequent modification of the training if indicated).

The front-end analysis (or training needs analysis) step in training design is critical. In particular, a comprehensive analysis of the tasks performed by equipment operators and maintainers is required before the training needs and associated functional specifications can be determined. The aim of the task analysis is to describe the knowledge, skills, and behaviours required for successful task performance, and identify the potential sources and consequences of human error. This task analysis would typically involve interviews with experts, reviews of written operating and maintenance procedures, and observations of equipment in use. It should include consideration of the information required by equipment operators and maintainers and how this information is obtained, the decision-making and problem-solving steps involved, the action sequences, and attentional requirements of the task. The task analysis should be conducted systematically, and well documented, to provide a solid foundation for the design of training and to provide a template for future training needs analyses.

An extension of the task analysis to include a cognitive task analysis may be justified for more complex task–equipment interfaces. Cognitive task analysis seeks to understand the cognitive processing requirements of task performance, typically through use of verbal protocols and structured interviews with experts. The outcomes of a cognitive task analysis include identification of the information used during complex decision-making, as well as the nature of the decision-making. The cognitive task analysis can also reveal information thatwill underpin the design of training and assessment. Again, the outcome of a cognitive task analysis may include identification of design deficiencies, which should be fed back into the design process.

The results of the task analysis are also used in the second phase of training design to define the actual contents of the training program, as well as the instructional strategy required. Regardless of the content of the training (the competencies required), or the methods employed, most effective instructional strategies embody four basic principles:

- The presentation of the concepts to be learned
- Demonstration of the knowledge, skills, and behaviours required
- Opportunities to practise
- Feedback during and after practise

An initial training design concept is typically refined iteratively through usability evaluation of prototype training models, until a fully functional final prototype is considered ready for full-scale development. Issues to be considered include the introduction of variation and the nature and scheduling of feedback. A compelling case has been presented to suggest that variation in the way tasks are ordered and in the versions of the tasks to be practised is important and that less frequent feedback should be provided. Whilst immediate performance may be reduced, retention and generalisation are enhanced as a consequence of the deeper information processing required during practise.

Evaluation of the consequences of training is also an essential and non-trivial step, and the task analysis aids in determining the appropriate performance measures to be used in evaluation (or competency assessment). A valid training evaluation requires careful selection of evaluation criteria and measures (closely connected to the task analysis results) and systematic collection and analysis of data. The use of simulation is a promising method for allowing trainees to be exposed to rare events, as well as for competency assessment.

7.3.2 Human Factors Engineering

"*Human factors engineering*" encompasses the consideration and quantification of human capabilities and limitations in system design, development, and evaluation (Horberry et al., 2011). In the automation and technology context, this is particularly important in the design of interfaces between people and automated components. While the use of human engineering standards (e.g., MIL-STD-1472H) may be useful, they are not sufficient. Prescriptive standards are often too general to be helpful in specific situations, they do not address tradeoffs that may be necessary, and they reflect the technology of the time at which they were written.

Other methods employed in human factors engineering include task analyses and human performance measures (e.g., workload, usability, situation awareness), as well as participatory human-centred design techniques (Horberry et al., 2018). Human-in-the-loop simulation allows analysis of the activities undertaken to achieve tasks during the design phase (INCOSE, 2023).

ISO 9241-210 *Ergonomics of human-system interaction Part 210: Human-centred design for interactive systems* (ISO, 2010b) provides principles for human-centred design of computer-based interactive systems that will be relevant to many technology projects, namely:

- The design is based upon an explicit understanding of users, tasks, and environments

The completed range of users and others who may be affected by the technology should be identified and considered. ISO 9241:210 suggests that a failure to understand adequately, user needs is a common source of system failure.

- Users are involved throughout design and development

Active involvement of users throughout the design process is critical. The nature and frequency of involvement will vary depending on the project however, the effectiveness of the user involvement will be proportional to the extent of direct interaction with technology designers.

- The design is driven and refined by user-centred evaluation

The risk of system failure is reduced by incorporating user feedback on preliminary designs into progressively refined solutions. User-centred evaluation is a key part of final acceptance testing, and ongoing feedback from users provides input into subsequent design improvements.

- The process is iterative

Effective utilisation of user feedback implies that multiple iterations of the design process will be required to progressively refine of specifications and prototypes.

- The design addresses the whole user experience

The effectiveness of the specific technology is only one aspect to be considered. Other aspects include the users' responses to the technology, as well as aspects of the implementation of the system such as training or wider impacts of the technology.

- The design team includes multidisciplinary skills and perspectives.

Achieving effective human-centred design requires diversity of skills within the design team and will likely require human factors expertise, in addition to engineering and software design. The involvement of users and other subject matter experts will be essential to achieving a satisfactory human-centred design.

Table 7.1 provides examples of the activities that comprise a human-centred design process and the outputs of each activity.

TABLE 7.1
Human-centred design activities (adapted from ISO 9241-210)

Activities	Detail	Outputs
1. Understand and specify the context of use	The characteristics of the users, tasks and organizational, technical and physical environment define the context in which the system is used.	Context of use description (eg user characteristics, tasks and goals, use environment).
2. Specify user requirements	User requirements provide the basis for the design and evaluation of systems to meet user needs. This includes user interface knowledge.	Context of use specification. User needs description and requirements specification
3. Produce design solutions to meet these requirements	Potential design solutions produced based on the context of use description, the state of the art in the domain, design guidelines, and the knowledge of the design team.	User interaction specification User interface specification Implemented user interface
4. Evaluate the designs against requirements	User-centred evaluation is a required activity at all HCD stages. Two widely used approaches are: inspection-based evaluation against usability guidelines, and user-based testing	Evaluation results Conformance test results Long-term monitoring results

Use-cases, that is, a description of a task performed by a person interacting with a system and the system responsibilities in accomplishing that task (Constantine & Lockwood, 2001), provide a starting point for user interface design.

Situation awareness refers to that portion of a person's knowledge pertaining to the state of a dynamic environment (Endsley, 1995). It is separate from decision-making and subsequent task performance. Poor decisions and/or task performance may still be made on the basis of accurate situation awareness; however, even the most highly trained and motivated operator will make poor decisions if their situation awareness is inaccurate or incomplete.

The first step in achieving accurate situation awareness is to perceive the state of relevant elements. This perception of the elements of the current situation is defined as level 1 situation awareness. The next step in the situation awareness process is for the operator to synthesise the level 1 elements into an understanding of the significance of those elements in the light of the operator's goals. Comprehension of the current situation is defined as level 2 situation awareness.

The final stage in the process is the prediction of the likely state of the situation in the near future. This projection of the future state is defined as level 3 situation awareness.

Loss of accurate situation awareness is a common cause of automation-related unwanted events. Interface design influences situation awareness by determining the information that can be acquired, the accuracy of the information, and the compatibility of the information with the operator's situation awareness needs (Endsley, 1995). Endsley and Jones (2012, 2024) have provided an approach to human-centred design focused on situation awareness that commences with describing situation awareness requirements through Goal-Directed Task Analysis. The goals and critical decisions associated with a job are described and the level 1, 2, and 3 situation awareness needs associated with these decisions are then determined.

The designers' challenge is then to provide interfaces that organise information around the user's situation assessment needs and assist the timely achievement of accurate level 3 situation awareness. General principles of designing for situation awareness have been developed by Endsley and Jones (2012, 2024), including:

- Organise information around goals
- Present Level 2 information directly – support comprehension
- Provide assistance for Level 3 situation awareness projections
- Support global situation awareness
- Explicitly identify missing information
- Support sensor reliability assessment
- Represent information timeliness
- Just say no to feature creep – buck the trend
- Insure logical consistency across modes and features
- Don't make people rely on alarms – provide projection support
- Make alarms unambiguous
- Reduce false alarms, reduce false alarms, reduce false alarms

- Set missed alarm and false alarm trade-offs appropriately
- Use multiple modalities to alarm, but insure they are consistent
- Keep the operator in control and in the loop
- Provide automation transparency
- Provide flexibility to support shared situation awareness across functions

7.3.3 Safety and Health

The "*safety*" domain includes consideration of safety risks such as those identified in ISO 17757. Relevant methods include traditional risk analysis and evaluation techniques such as hazard and operability studies, layers of protection analysis, failure modes and effects analysis, as well as functional safety analyses, and systems-theoretic process analysis (STPA).

STPA, in particular, may be useful for analysis of complex systems involving automated components because both software and human operators are included in the analysis. STPA is a proactive analysis method that identifies potential unsafe conditions during development and avoids the simplistic linear causality assumptions inherent in traditional techniques. Safety is treated as a control problem rather than a failure prevention problem. Unsafe conditions are viewed as a consequence of complex dynamic processes that may operate concurrently. STPA also includes consideration of the wider, dynamic, organisational context in which the automated system is situated. STPA has been successfully used during the development of automated bulldozers and automated haulage. Other systems-based analysis techniques (e.g., SAfER) may also be useful (Hassall et al., 2014).

The "*occupational health*" domain encompasses the use of risk management techniques, and task-based risk assessment (Burgess-Limerick et al., 2012) to ensure that the system design minimises risks of adverse health consequences to system operators and maintainers, and indeed, anyone else potentially impacted by the system activities. These analyses should encompass all operational and maintenance activities associated with the autonomous component or system.

One health issue associated with the introduction of autonomous systems to mining is the potential impact on the physical and mental health of control-room operators tasked with interacting with autonomous systems. Stress associated with high (or low) cognitive workloads, potentially combined with reduced social interactions and low control of workload, and/or production pressures, may lead to adverse mental health consequences.

An overall focus on human-systems integration includes consideration of interactions and potential trade-offs between decisions made in different domains. For example, decisions regarding automation and interface complexity may influence personnel characteristics and training requirements, as well as the anticipated number of people required for system operation and maintenance; while issues highlighted during the safety analysis may well lead to additional human factors engineering work to reduce risks.

7.4 IMPLEMENTATION OF HUMAN SYSTEMS INTEGRATION IN MINING

Guidance provided for the rail industry (Melnik et al., 2018) has been adapted here for the acquisition of new mining technologies. Although the stages of systems engineering are presented sequentially, the reality is that iterative loops occur both within stages and between stages (Folds, 2015). While the results of evaluations conducted during design and development will certainly influence subsequent design iterations, they may also feedback to changes to requirements, or even result in changes to the concept of operations.

7.4.1 Analysis

The initial stage of the systems engineering process is analysis. Human-centred analysis activities conducted as part of human systems integration address the following:

- Concept of operation – What are the goals of the system and, in particular, what are the anticipated operational and maintenance roles that people will play? Who will these people be? What knowledge and skills will they have? What diversity is anticipated? Are there other people inside or outside the system that should be considered?
- Contexts – What is the range of operational contexts and use cases? Are there different modes of operation? What range of environmental conditions is anticipated?
- Tasks – how will functions be allocated within the system? What physical tasks will people need to perform? What monitoring or decision-making tasks need to be undertaken? What current tasks will no longer be undertaken or altered? What are the critical tasks that are performed by people? A variety of task analysis techniques may employed depending on the nature of the tasks. Similarly, analyses of workload and situation awareness are likely to be appropriate.
- Known challenges/lessons learned – Are there known human performance concerns based on experiences with similar systems in the same or other industries? What can be learned from previous incidents or near-misses?
- Safety and health – What hazards may be present? How could adverse safety or health outcomes occur? What errors could people make and what would be the consequences? How can the potential for detection of both human and technological errors, and recovery from errors, be increased? What critical controls are required to prevent or mitigate adverse safety or health outcomes?
- Tradeoffs – Are there tradeoffs between human-systems integration domains that need to be evaluated? Are there tradeoffs between the human-systems integration domains and other systems engineering elements (e.g., cost) that require examination?

7.4.2 Requirements

The output of these analyses leads to human-systems integration requirements that inform subsequent system design and development. Potential requirements include:

- Information – What information needs to be received by people in the system to maintain situation awareness? How should the information be presented to best support decision-making?
- Control – What controls and modes of interaction with the system are required?
- Communication – What communication channels are required inside and outside the system? What methods of communication should be provided?
- Physical environment – What physical workstation designs are required, e.g., layout, lighting, visibility, reachability? How will human diversity be accommodated?
- Selection and Training – How will the people in the system be selected? What training (initial and ongoing) will be required? How should the training be undertaken? How will competency be assessed and reassessed?

7.4.3 Design

Based the explicit understanding of users, tasks, and environments, a human-centred design and development process involving users is undertaken by a multidisciplinary team including human factors expertise. The process is iterative, likely involving the design and testing of prototypes of increasing fidelity, and likely to involve human-in-the-loop simulation.

Design and development outcomes will include:

- Work environment – Design of physical environments to maximise performance, as well as health and safety. Human engineering standards may be particularly relevant to physical design.
- Software and interfaces – Design of the overall software architecture, as well as the interfaces through which information is received by humans, and through which input is given by humans, to ensure efficient and safe performance under normal and abnormal conditions.
- Training – Design of the curriculum, training methods, and competency assessments.
- Documentation – Developing readable, understandable, and usable procedures, training manuals, and related operations and maintenance documentation that reflect "work-as-done" rather than "work-as-imagined".

7.4.4 Testing and Evaluation

User-centred evaluation occurs throughout the entire systems engineering process, as well as at final system validation. Testing and evaluation activities include:

- Planning – Human systems integration issues should be incorporated into the overall systems engineering testing and evaluation program.
- Evaluation of prototypes – Users representing the diversity of the intended workforce participate in evaluations of prototypes of increasing fidelity. Both physical and virtual simulations may be useful, human-in-the-loop simulation even more so.
- Human engineering discrepancy resolution – Aspects of the design that do not meet requirements during the iterative evaluations are systematically identified and tracked. Corrective actions are proposed and implemented.
- Final validation – Each requirement requires evaluation in the final system validation. Evaluation scenarios include the contexts and use cases identified during the analysis stage. Data collected will include process measures (e.g., workload and situation awareness) and outcome measures, as well as user evaluations.

7.4.5 Human Systems Integration Program Plan

During the preparation of proposals to implement any new technology at mines, and particularly if automated components are involved, vendors should be required to submit a human systems integration program plan that details the human-systems integration work that will be performed in collaboration with the purchaser; how it will be done; and by whom.

A human systems integration program plan should include:

- Overview – An overview of the proposed system; preliminary concept of operations, associated human roles, and operational environment; experiences with predecessor systems.
- Organisational capabilities – Summary job descriptions and the qualifications of key human systems integration practitioners within the vendor.
- Program Risks – A discussion of how human systems integration risks will be identified and addressed.
- Human systems integration activities – The specific human-systems integration activities that will be performed by the vendor in collaboration with the purchaser to address each of the domains of human-systems integration during system analysis, design, and evaluation. Identification of who will undertake these activities.
- Human systems integration schedule – A milestone chart identifying each human systems integration activity, including key decision points, and their relationship to the programme milestones.

7.5 HUMAN READINESS LEVELS

Judging the suitability of any new technology for deployment depends on an assurance that both that the technology will function as intended, and that the use of the technology by humans in the system will have the intended outcome. As mentioned earlier in this book, technology readiness levels are commonly used to describe the development of technology. Human readiness levels are an analogous scale used to evaluate, track, and communicate the preparedness of a technology for human use. HRL have been formalised through HFES/ANSI 400-2021 (HFES/ANSI, 2021). The concordance between technology readiness levels and human readiness levels is illustrated in Table 7.2. Achieving satisfactory human readiness levels requires a human-centred design process during technology development.

Annex C of HFES/ANSI 400-2021 provides questions to serve as triggers to ensure that critical human systems evaluations are not omitted at each level.

Questions for HRL level 1 include:

- Have preliminary usage scenarios for potential users been identified?
- Have potential key human performance issues and risks been identified and concomitant basic research conducted?

For HRL level 2 the questions include:

TABLE 7.2
Technological and Human Readiness Levels compared

Level	Technology Readiness Level	Human Readiness Level
1	Basic principles observed	Basic principles for human characteristics, performance, and behaviour observed and reported
2	Technology concept formulated	Human-centred concepts, applications, and guidelines defined
3	Experimental proof of concept	Human-centred requirements to support human performance and human-technology interactions established
4	Technology validated in laboratory	Modelling, part-task testing, and trade studies of human systems design concepts and applications completed
5	Technology validated in relevant environment	Human-centred evaluation of prototypes in mission-relevant part-task simulations completed to inform design
6	Technology demonstrated in relevant environment	Human systems design fully matured and demonstrated in a relevant high-fidelity, simulated environment or actual environment
7	System prototype demonstration in operational environment	Human systems design fully tested and verified in operational environment with system hardware and software and representative users
8	System complete and qualified	Human systems design fully tested, verified, and approved in mission operations, using completed system hardware and software and representative users
9	System proven on operational environment	System successfully used in operations across the operational envelope with systematic monitoring of human system performance

- Have key human-centred design principles, standards, and guidance been established?
- Has human performance on legacy or comparable systems been analysed to understand key human-technology interactions, human behaviour, and human performance issues?
- Have potential sources of human error and misuse been identified?

For HRL level 3 the questions include:

- Have human systems experts with requisite expertise been engaged and funded to support the design and development effort?
- Have cognitive task analyses and function and task analyses for each user role been completed?
- Have situation awareness information flow and sharing requirements across teams of human or automated system components been identified?
- Have initial safety analyses for human users been completed?
- Have initial manpower, personnel, and training analyses been completed?

For HRL level 4 the questions include:

- Have usage scenarios been updated, based on modelling and part-task testing?
- Have strategies to mitigate safety implications for human users been identified and recommended?
- Have strategies to accommodate manpower, personnel, and training concerns been identified and recommended?
- Has conformance of preliminary designs to human performance requirements, design principles, standards, and guidance been verified?

For HRL level 5 the questions include:

- Have functioning prototypes of the human-system interface and simulations of mission tasks and conditions been developed to support assessment of critical human performance issues?
- Have task analyses been updated, based on prototype testing in mission-relevant part-task simulations?
- Has the suitability of human-machine teaming strategies and human-machine function allocations been determined, based on prototype testing in mission-relevant part-task simulations?

For HRL level 6 the questions include:

- Has the full range of user scenarios and tasks been tested in high-fidelity simulated or actual environments?

- Has a system to track and resolve human-systems issues after fielding been developed and evaluated in high-fidelity simulated or actual environments?
- Have relevant human performance data been collected and evaluated to determine whether human performance metrics are successfully met, based on testing in high-fidelity simulated or actual environments?
- Has conformance of functional prototypes to human performance requirements, design principles, standards, and guidance been verified?

For HRL level 7 the questions include:

- Has the effectiveness of strategies to mitigate safety implications for human users been evaluated with the final development system in an operational environment?
- Have human user procedures been tested with the final development system in an operational environment?
- Have relevant human performance data been collected and evaluated to determine whether human performance metrics are successfully met, based on testing with the final development system in an operational environment?
- Has conformance of the final development system to human performance requirements, design principles, standards, and guidance been verified?

For HRL level 8 the questions are:

- Has the effectiveness of strategies to accommodate manpower, personnel, and training concerns been evaluated and successfully demonstrated with the production system in mission operations?
- Have relevant human performance data been evaluated to determine whether human performance metrics are successfully met, based on qualification of the production system in mission operations?
- Have human use issues been satisfactorily resolved, as evidenced by qualification of the production system in mission operations?

For HRL level 9 the questions include:

- Are human systems performance data and lessons learned being documented for recommended systems improvements and future applications?
- Does the system designed to track and resolve human-systems issues in the fielded system fully support these activities?
- Is user training for operation of the fielded system being evaluated for required modifications?
- Are potential upgrades to the fielded system being evaluated to address human-systems issues and impacts?

Handley (2023) suggests that data to evaluate the criteria at each human readiness level can be obtained by completing activities identified by the Human System

Integration Framework (HSIF), a software tool developed by the US Air Force (Lacson et al., 2017).

> The HSIF tool looks a bit like a flowchart, a scrollable canvas that covers the entire acquisition cycle, from early development planning to operation and sustainment to disposal. The acquisition timeline is listed across the top of the canvas, and the HSI integrator roles and domains are listed down the left side. The rest of the canvas is filled with task boxes, many linked together to show collaboration, that detail specific HSI activities for each stage. When a task box is selected, it expands to show a description of the activity, hyperlinked references to consult for more information and products that might result from the activity. The expanded view also includes a place for comments about HSI risks and trade-offs, as well as a tab to document deliverables and more. The tool is completely searchable, and the reference section includes an acronym finder, a product library and a database of all the hyperlinked references in the tool, more than 200 documents in all. Aside from the basic features and information, users can flag certain task boxes, assign due dates, run reports, save customized versions of the canvas and export content to a spreadsheet, for example.
>
> **(Bowden, 2016)**

The HSIF software also serves as a mechanism for sharing information during procurement and as a record management tool.

7.6 CONCLUSIONS

While the introduction of automated mining equipment has great potential to reduce safety and health risks, new credible failure modes are introduced. The new failure modes all have human aspects. Current standards and guidance materials pay insufficient attention to the integration of humans and technology during the implementation of automation in mining.

Issues of particular importance include the design of interfaces to maintain situation awareness, the reduction of control room operator workloads, and the training of people who will undertake new roles. The extent of training required for all those impacted by the technology should not be underestimated and will likely be increased compared to previous roles. Ongoing training and competency assessment will be required as the systems are modified. Ensuring that sufficient numbers of trained control room staff are available to the industry is critical for both productivity and safety and health.

Human systems integration processes should be implemented during acquisition of automated mining equipment, and technology vendors should be required to provide a human systems integration program plan. This book has been intended to provide guidance in developing and executing such a plan.

References

Abbott, P.A., & Weinger, M.B. (2020). Health information technology: Fallacies and sober realities–Redux A homage to Bentzi Karsh and Robert Wears. *Applied Ergonomics, 82*, 102973.

Albus, J. Quintero, R., Huang, H-M., & Roche, M. (1989). *Mining Automation Real-time Control System Architecture Standard Reference Model (MASREM)*, vol. 1. NIST Technical Note 1261.

Androulakis, V., Sottile, J., Schafrik, S., & Agioutantis, Z. (2019). Elements of autonomous shuttle car operation in underground coal mines. Proceedings of the 2019 IEEE Industry Applications Society Annual Meeting, pp. 1–7.

Atkinson, G. (1996). Automated mining machine safety investigation. Unpublished MSc thesis. McGill University.

Australian Transport Safety Bureau. (2022). Runaway and derailment of loaded ore train M02712. https://www.atsb.gov.au/publications/investigation_reports/2018/rair/ro-2018-018 (accessed October 14, 2024)

Baggermann, S., Berdich, D., & Whitmore, M. (2009). Human systems integration (HSI) case studies from NASA constellation program. Paper presented at the Human Systems Integration. https://ntrs.nasa.gov/api/citations/20090001318/downloads/20090001318.pdf

Bellamy, D., & Pravica, L. (2011). Assessing the impact of driverless haul trucks in Australian surface mining. *Resources Policy, 36*, 149–158.

Benlaajili, S., Moutaouakkil, F., & Chebak, A. (2021). Infrastructural requirements for the implementation of autonomous trucks in open-pit mines. Proceedings of the VIth International Innovative Mining Symposium E3S Web of Conferences 315, 03009.

Bhattacharya, J. (2020). Wireless network capacity and capability is a pre-requirement for implementation of automation and other technologies in open-pit mining. *Journal of Mines, Metals and Fuels, 68,* 152–152.

BHP (2022a). BHP unveils world-first automated shiploaders. https://www.bhp.com/news/media-centre/releases/2022/06/bhp-unveils-world-first-automated-shiploaders (accessed June 7, 2023).

Boloz, L., & Bialy, W. (2020). Automation and robotisation of underground mining in Poland. *Applied Sciences, 10*, 7221.

Bowden, K. (2016). Human systems integration framework tool improves process for system acquisition professionals. Wright Patterson Airforce Base News. https://www.wpafb.af.mil/News/Article-Display/Article/937334/human-systems-integration-framework-tool-improves-process-for-system- acquisitio/ (accessed October 14, 2024).

Booher, H. (2003). *Handbook of Human Systems Integration*. New York: Wiley.

Brodny, J., & Tutak, M. (2018). Exposure to harmful dusts on fully powered longwall coal mines in Poland. *International Journal of Environmental Research and Public Health, 15,* 1846.

Burger, D.J. (2006). Integration of the mining plan in a mining automation system using state-of-the-art technology and De Beers Finch mine. *The Journal of the South African Institute of Mining and Metallurgy, 106*, 553–559.

Burgess-Limerick, R. (2020). Human-systems integration for the safe implementation of automation. *Mining, Metallurgy & Exploration, 37*, 1799–1806.

Burgess-Limerick, R., Horberry, T., Cronin, J., & Steiner, L. (2017). Mining automation human-systems integration: A Case study of success at CMOC-Northparkes. Proceedings of the 13th AusIMM Underground Operators' Conference. AusIMM, Melbourne, pp 93–98.

Burgess-Limerick, R., Joy, J., Cooke, T., & Horberry. T (2012). EDEEP - An innovative process for improving the safety of mining equipment. *Minerals, 2*, 272–282.

Candappa, N. (2020). An investigation of fatal and serious injury wire rope safety barrier crashes using crash data and software. Unpublished PhD thesis, Monash University, Australia.

Chan, R. (2022). Safety-first approach drives remote dozer solution. *Australian Mining*. https://www.australianmining.com.au/safety-first-approach-drives-remote-dozer-solution/ (accessed June 15, 2023)

Chirgwin, P. (2021a). Skills development and training of future workers in mining automation control rooms. *Computers in Human Behavior Reports, 4*, 100115.

Chirgwin, P. (2021b). In control with higher education through work-based learning. *Journal of Higher Education Theory and Practice, 21*(15), 43–52.

Cholteeva, Y. (2021). Autonomous light vehicles: How driverless technology is trickling down through Australian mining. *Mining Technology*. https://www.mining-technology.com/features/autonomous-light-vehicles-how-driverless-technology-is-trickling-down-through-australian-mining/ (accessed June 7, 2023).

Christofidis, M. J., Hill, A., Horswill, M. S., & Watson, M. O. (2016). Observation chart design features affect the detection of patient deterioration: a systematic experimental evaluation. *Journal of Advanced Nursing*, 72(1), 158-172.

Corke, P., Roberts, J., Cunningham, J., & Hainsworth, D. (2008). Mining robotics. Chapter 49 In B. Siciliano & O. Khatib (Eds.), *Springer Handbook of Robotics*, pp. 1127–1150. New York: Springer.

Corke, P., Roberts, J., & Winstanley, G. (1998). Vision-based control for mining automation. *IEEE Robotics & Automation Magazine*, December, 44–49.

Cornish, L., Hill, A., Horswill, M. S., Becker, S. I., & Watson, M. O. (2019). Eye-tracking reveals how observation chart design features affect the detection of patient deterioration: An experimental study. *Applied Ergonomics, 75*, 230–242.

Craig, B. (2022). Western Australia Iron Ore Update. *Presentation & Speech*, October 3. https://www.bhp.com/-/media/documents/media/reports-and-presentations/2022/221003_waiospeeches.pdf (accessed June 3, 2023).

Cressman, T.J. (2023). Assisting operators of tomorrow. Proceedings of the 26th World Mining Congress. Brisbane, June, pp. 283–294.

Cummings, M.L. (2023a) Revising human-systems engineering principles for embedded AI applications. *Frontiers of Neuroergonomics, 4*, 1102165.

Cummings, M.L. (2023b). What self-driving cars tell us about AI risks. *IEEE Spectrum*. https://spectrum.ieee.org/self-driving-cars-2662494269 (accessed July 31, 2023).

Dekker, S., Cilliers, P., & Hofmeyr, J.-H. (2011). The complexity of failure: Implications of complexity theory for safety investigations. Safety Science, 49(6), 939-945.

Department of Mines and Petroleum - Resources Safety. (2015). Significant Incident Report No. 226 - Subject: Collision between an autonomous haul truck. https://www.dmp.wa.gov.au/Documents/Safety/MS_SIR_226_Collision_between_an_autonomous_haul_truck_and_manned_water_cart.pdf and manned water cart. (accessed October 14, 2024)

Design Controls, 21 C.F.R. § 820.30 (2021). https://www.ecfr.gov/current/title-21 (accessed August 8, 2022).

Dragt, B.J., Camisani-Calzolari, F.R., & Craig, I.K. (2005). An overview of the automation of load-haul-dump vehicles in an underground mining environment. Proceedings of the International Federation of Automatic Control, 16th Triennial World Congress. Prague, pp. 37–48.

Du Venage, G. (2019). Fully automated sub level caving goes live in Mali. *Engineering and Mining Journal, 220*(1), 38.

Dudley, J.J. (2014). Enhancing awareness to support teleoperation of a bulldozer. MPhil Thesis, School of Mechanical and Mining Engineering, The University of Queensland. Australia.

Dudley, J.J., Aw, J., & McAree, P.R. (2013). *Minimal Perception Requirements to Support Effective Remote Control of Bulldozers* (ACARP Project C20021). Brisbane Qld Australia: ACARP.

Dudley, J.J., & McAree, R. (2016). *Shovel Load Assist Project - ACARP Project Report C16031.* Brisbane Qld Australia: ACARP (Australian Coal Industry Research Program)

Duff., E., Roberts, J., & Corke, P.I. (2002). Automation of an underground mining vehicle using reactive navigation and opportunistic localisation. Proceedings of the 2002 Australasian Conference on Robotics and Automation, pp. 151–156.

Dunbabin, M., & Corke, P. (2006). Autonomous excavation using a rope shovel. In P. Corke & S. Sukkariah (Eds.), *Field and Service Robotics. Springer Tracts in Advanced Robotics,* vol. 25. Berlin, Heidelberg: Springer.

Dunn, M.T., Reid, P.B., Thompson, J.P., & Beyers, J.G. (2023). Applications of advanced sensors and digital platforms for undergound mining automation. Proceedings of the 26th World Mining Congress. Brisbane, June, pp. 306–320.

Dunn, M.T., Reid, D., & Ralston, J. (2015). Control of automated mining machinery using aided inertial navigation. In J. Billingsley & P. Brett (Eds.), *Machine Vision and Mechatronics in Practice*, pp. 1–9. Berlin, Heidelberg: Springer.

Dyson, N. (2020a). Underground automation leader reaps benefits. *Australian Mining Monthly.* https://www.miningmonthly.com/technology-innovation/news/1378395/underground-automation-leader-reaps-benefits (accessed June 8, 2023).

Dyson, N. (2020b). Syama's automation surge. *Mining Magazine.* https://www.miningmagazine.com/technology-innovation/news/1387604/syama%E2%80%99s-automation-surge (accessed August 6, 2023).

Ellis, Z. (2023). Autonomous surface drilling: KGHM Robinson mine. Presentation to the Mine Automation and Emerging Technologies Health and Safety Partnership Meeting, September 21.

Endsley, M. (1995). Toward a Theory of Situation Awareness in Dynamic Systems. Human Factors, 37, 32-64.

Endsley, M. R., & Jones, D. G. (2012). Designing for situation awareness: An approach to human-centered design (2nd ed.). Taylor & Francis.

Endsley, M. R. & Jones, D. G. (2024). Situation Awareness Oriented Design: Review and Future Directions. *International Journal of Human–Computer Interaction.* https://doi.org/10.1080/10447318.2024.2318884.

European Organisation for the Safety of Air Navigation (Eurocontrol) (2019). *Human Factors Integration in ATM System Design.* Brussels: EuroControl.

Federal Railroad Administration (FRA) (2019). Human Systems Integration: What is HSI? And why does it matter to the rail industry? https://railroads.dot.gov/human-factors/elearning-attention/human-systems-integration

Fisher, B.S., & Schnittger, S. (2012). *Autonomous and Remote Operation Technologies in the Mining Industry: Benefits and Costs.* BAE Research Report 12.1. Canberra: BAEconomics Pty Ltd.

Flach, J. M., Jacques, P. F., Patrick, D. L., Amelink, M., Van Paassen, M. M., & Mulder, M. (2003). A search for meaning: A case study of the approach-to-landing. *Handbook of Cognitive Task Design*. Boca Raton FL: CRC Press.171-191.

Folds, D. (2015). Systems engineering perspective on human systems integration. In D. Boehm-Davis, F. Durso, & J. Lee (Eds.), *APA Handbook of Human Systems Integration*. Washington, DC: American Psychological Association.

Food and Drug Administration. (2016a). Applying human factors and usability engineering to medical devices: Guidance for industry and food and drug administration staff. https://www.fda.gov/regulatory-information/search-fda-guidance-documents/applying-human-factors-and-usability-engineering-medical-devices (accessed August 8, 2022).

Food and Drug Administration. (2016b). Design control guidance for medical device manufacturers. https://www.fda.gov/regulatory-information/search-fda-guidance-documents/design-control-guidance-medical-device-manufacturers (accessed August 8, 2022).

Fouche, L. (2023). Vehicle fatality elimination. Presentation to the Collision Avoidance Forum 2023. https://www.resourcesregulator.nsw.gov.au/sites/default/files/2023-03/Leon-Fouche-Rio-Tinto-Update-CA-Forum.pdf (accessed June 3, 2023).

FutureBridge (2022). Autonomous haulage systems – The future of mining operations. https://www.futurebridge.com/industry/perspectives-industrial-manufacturing/autonomous-haulage-systems-the-future-of-mining-operations/ (accessed May 24, 2023).

Gaber, T., El Jazouli, Y., Eldesouky, E., & Ali, A. (2021). Autonomous haulage systems in the mining industry: Cybersecurity, communication and safety issues and challenges. *Electronics, 10*(11), 1357.

Ghodrati, B., Hoseinie, S.H., & Garmabaki, A.H.S. (2015). Reliability considerations in automated mining systems. *International Journal of Mining, Reclamation and Environment, 29*, 404–418.

Gleeson, D. (2021a). Glencore showcases automated longwall advancements at Oaky Creek. *International Mining*. https://im-mining.com/2021/09/30/glencore-showcases-automated-longwall-advancements-oaky-creek/ (accessed June 7, 2023).

Gleeson, D. (2021b). Thiess hits new heights with SATS dozer technology at Lake Vermont. *International Mining*. https://im-mining.com/2021/06/02/thiess-hits-new-heights-sats-dozer-technology-lake-vermont/ (accessed August 9, 2023).

Gleeson, D. (2022). Anglo American's longwall automation milestone recognised in awards ceremony. *International Mining*. https://im-mining.com/2022/09/02/anglo-americans-longwall-automation-milestone-recognised-in-awards-ceremony/ (accessed June 7, 2023).

Gleeson, D. (2023). Huawei boosts Shaanxi Coal production efficiency & safety with 5G-backed solution. *International Mining*. https://im-mining.com/2023/05/18/huawei-boosts-shaanxi-coal-production-efficiency-and-safety-with-5g-backed-solution/ (accessed July 17, 2023).

GMG (2019). *Guideline for the Implementation of Autonomous Systems in Mining*. https://gmggroup.org/guideline-for-the-implementation-of-autonomous-systems-in-mining-2/ (accessed October 14, 2024).

GMG (2021). Rio Tinto's experience with automation improving safety for employees and creating value. https://gmggroup.org/wp-content/uploads/2021/03/2021-01-11-Rio-Tintos-Experience-with-Automation-and-People.pdf (accessed June 7, 2023).

Gordon, S.E. (1994). *Systematic Training Program Design: Maximising Effectiveness and Minimizing Liability*. Englewood Cliffs, NJ: Prentice Hall.

Hamburger, P.S. (2008). Ten questions: An interview with Patricia S. Hamburger, Director, human systems integration engineering, Naval Sea Systems Command (NAVSEA). *Naval Engineers Journal, 120*, 15–21.

Handley, H. (2023). Human system integration framework (HSIF) activities to support human readiness levels (HRLs). *Ergonomics in Design*. https://doi.org/10.1177/10648046231152050

Hariyadi, A., Castro, M., & Fisher, J. (2016). Automated train ore transport. In *Proceedings of the Seventh International Conference & Exhibition on Mass Mining (MassMin 2016)*, pp. 563–570. Melbourne: AusIMM.https://www.ausimm.com/publications/conference-proceedings/seventh-international-conference--exhibition-on-mass-mining-massmin-2016/

Hassall, M. E. (2013). Methods and tools to help industry personnel identify and manage hazardous situations. (Doctor of Philosophy). The University of Queensland, Queensland, Australia.

Hassall, M.E., Sanderson, P.P., & Cameron, I.T. (2014). The development and testing of SAfER: A resilience-based human factors method. *Journal of Cognitive Engineering and Decision Making, 8,* 162–186.

Hassall, M. E., Joy, J., Doran, C., & Punch, M. (2015). Selection and optimisation of risk controls (ACARP report C23007).

Hassall, M.E., Seligmann, B., Lynas, D., Haight, J., & Burgess-Limerick, R. (2022). Predicting human-system interaction risks associated with autonomous systems in mining. *Human Factors in Robots, Drones and Unmanned Systems, 57,* 78–85.

HFES/ANSI. (2021). Human readiness level scale in the system development process (HFES/ANSI 400-2021). https://www.hfes.org/publications/technical-standards

Horberry, T. (2012). The health and safety benefits of new technologies in mining: A review and strategy for designing and deploying effective user-centred systems. *Minerals, 2,* 417–425.

Horberry, T., Burgess-Limerick, R., Cooke, T., & Steiner, L. (2016). Improving mining equipment safety through human-centered design. *Ergonomics in Design, 24*(3), 29–34. https://doi.org/10.1177/1064804616636299

Horberry, T., Burgess-Limerick, R., & Steiner, L. (2011). *Human Factors for the Design, Operation and Maintenance of Mining Equipment.* Boca Raton: CRC Press.

Horberry, T., Burgess-Limerick, R., & Steiner, L. (2018). *Human-Centered Design for Mining Equipment and New Technology.* Boca Raton: CRC Press.

Horberry, T., Mulvihill, C., Fitzharris, M., Lawrence, B., Lenne, M., Kuo, J., & Wood, D. (2021). Human-centered design for an in-vehicle truck driver fatigue and distraction warning system. *IEEE Transactions on Intelligent Transport Systems.* https://doi.org.10.1109/TITS.2021.3053096

Horberry, T., Harris, J., Cliff, D., Dodshon, P., Lee, J., Lim, N., & Way, K. (2024). Fatigue management in mining: current practices and future directions. *Journal of Work Health and Safety Regulation* 23-004 . https://doi.org/10.57523/jaohlev.oa.23-004

ICMM (2015). *Critical Control Management: Implementation Guide.* London: International Council of Mining and Metals.

International Council on Systems Engineering (2015). *INCOSE Systems Engineering Handbook: A Guide for System Life Cycle Processes and Activities.* Hoboken, NJ: Wiley.

International Council on Systems Engineering (2023). *Human Systems Integration: A Primer. Volume 1.* INCOSE. https://www.incose.org/publications/products/hsi-primer (accessed October 14, 2024).

International Electrotechnical Commission. (2018). *Risk management - Risk assessment techniques.* ISO 31010: Geneva: International Electrotechnical Commission

International Standards Organisation (2010a). *Safety of Machinery — General Principles for Design — Risk Assessment and Risk Reduction.* ISO 12100: 2010. Geneva: ISO

International Standards Organisation (2010b). *Ergonomics of Human-system Interaction Part 210: Human-centred Design for Interactive Systems.* ISO9241-210. p. 9. Geneva: ISO

International Organization for Standardization (2019). *Ergonomics of human-system interaction—Part 210: Human-centred design for interactive systems.* ISO 9241-210. https://www.iso.org/standard/77520.html

International Standards Organisation. (2018). ISO 31000: *Risk management - Guidelines.* Geneva: ISO

International Organization for Standardization (2019). *Ergonomics of human-system interaction—Part 210: Human-centred design for interactive systems.* ISO 9241-210. https://www.iso.org/standard/77520.html

Ishimoto, H., & Hamada, T. (2020). Safety concept and architecture for autonomous haulage system in mining. Proceedings of the 37th International Symposium on Automation and Robotics in Construction, pp. 377–384.

Khashe, Y., & Meshkati, N. (2019). Human and organizational factors of positive train control safety system the application of high reliability organizing in railroad. *Proceedings of the Human Factors and Ergonomics Society Annual Meeting, 63*(1), 1824–1828. https://doi.org/10.1177/1071181319631324

Knights, P., & Yeates, G. (2021). Progress toward zero entry mining: Automation enabling safer, more efficient mining. *IEEE Industrial Electronics Magazine, 15*, 32–38.

Konnoth, D. (2022). Are electronic health records medical devices? In I.G. Cohen, T. Minssen, W.N. Price, C. Robertson, & C. Shachar (Eds.), *The Future of Medical Device Regulation: Innovation and Protection,* pp. 36–46. Cambridge: Cambridge University Press.

Lacson, F.C., Risser, M.R., & Gwynne, J.W. (2017). The Human Systems Integration Framework: Enhanced HSI support for system acquisition. Proceedings of the Human Factors and Ergonomics Society 2017 Annual Meeting.1720-1724.

LaTourrette, T., & Regan, L. (2022). *Barriers to the Commercialisation and Adoption of New Underground Coal Mining Technologies in the U.S.* Interim Results. RAND Corporation. https://www.rand.org/pubs/working_papers/WRA1575-1.html (accessed August 8, 2023).

Leeming, J.J. (2023). Full automation of resin roofbolting in soft rock mining. Proceedings of the 26th World Mining Congress. Brisbane, June 2023, pp. 405–414.

Leonida, C. (2023). Dig, Load, Haul — Repeat. *Engineering & Mining Journal.* https://www.e-mj.com/features/dig-load-haul-repeat/ (accessed June 8, 2023).

Lever, P. (2011). Automation and robotics. In *SME Mining Engineering Handbook* (3rd ed., pp. 805–824). Society for Mining, Metallurgy and Exploration.

Leveson, N. G. (2004). *A new accident model for engineering safer systems.* Safety Science, 42(4), 237-270.

Leveson, N.G. (2012). *Engineering a Safer World: Systems Thinking Applied to Safety.* Boston: MIT Press.

Leveson, N.G., & Thomas, J.P. (2018). *STPA Handbook.* Boston: MIT Press.

Li, J., & Shan, K. (2018). Intelligent mining technology for an underground metal mine based on unmanned equipment. *Engineering, 4*, 381–391.

Liu, K., Valerdi, R., Rhodes, D., Kimm, L., & Headen, A. (2010). *The F119 Engine: A Success Story of Human Systems Integration in Acquisition.* Defense Acquisition University. https://apps.dtic.mil/sti/pdfs/ADA518530.pdf

Lynas, D., & Burgess-Limerick, R. (2019). Whole-body vibration associated with dozer operation at an Australian surface coal mine. *Annals of Work Exposures and Health, 63*, 881–889.

Lynas, D., & Horberry, T. (2011). Human factor issues with automated mining equipment. *The Ergonomics Open Journal, 4*(Suppl 2-M3), 74–80.

Marshall, J.A., Bronchi's, A., Reboot, E., & Scheming, S. (2016). Robotics in mining. Chapter 59 In B. Siciliano & O. Khatib (Eds.), *Springer Handbook of Robotics* (2nd ed., pp. 1549–1576). New York: Springer

McAree. R., Hensel, R., & Smith, Z. (2017). *Automated Bulk Dozer Push - Reducing the Cost of Overburden Removal.* (ACARP C24037) Brisbane Qld Australia: ACARP.

McAree, R. (2022). Personal communication, November 2022.

Meers, L. Van Duin, S., & Ryan, M. (2013). Rapid roadway development. Proceedings of the 30th International Symposium on Automation and Robotics in Construction and Mining; Held in conjunction with the 23rd World Mining Congress, Montreal.

Melnik, G., Roth, E., Multer, J., Safar, H., & Isaacs, M. (2018). *An Acquisition Approach to Adopting human Systems Integration in the Railroad Industry.* DOT/FRA/ORD-18/05. Washington DC: US Department of Transport.

Moore, P. (2023). Komatsu starts commercial teleop of large ICT mining dozers at Anglo's Minas-Rio. *International Mining.* https://im-mining.com/2023/06/14/komatsu-starts-commercial-teleop-of-large-ict-mining-dozers-at-anglos-minas-rio/ (accessed June 15, 2023).

Moreau, K., Lamanen, C., Bose, R., Shang, H., & Scott, J.A. (2021). Environmental impact improvements due to introducing automation into underground copper mines. *International Journal of Mining Science and Technology, 31*, 1159–1167.

Morton, J. (2017). New blast hole drill rigs feature more autonomy. *Engineering and Mining Journal*, March, 28–35. https://www.e-mj.com/features/new-blasthole-drill-rigs-feature-more-autonomy/ (accessed October 14, 2024).

National Transport Safety Board. (2019). Collision Between Vehicle Controlled by Developmental Automated Driving System and Pedestrian - Tempe, Arizona, March 18, 2018. Washington DC: NTSB.

Nebot, E. (2007). Surface mining: Main research issues for autonomous operation. *Robotics Research, Springer Transactions in Advanced Robotics, 28*, 268–280.

NSW Resources Regulator (2019). Collision between semi-autonomous dozer and an excavator. DOC19/758086. Sydney: Department of Planning, Industry and Environment.

Ninness, J. (2018). Autonomous mining trucks | Are there any limitations? *Australasian Mine Safety Journal*, November 13. https://www.amsj.com.au/autonomous-mining-truck-safety/ (accessed May 24, 2023).

Onifade M., Said, K.O., & Shivute, A.P. (2023). Safe mining operations through technological advancement. *Process Safety and Environmental Protection, 175*, 251–258

Paraszczak, J., Gustafson, A., & Schunnesson, H. (2015) Technical and operational aspects of autonomous LHD application in metal mines. *International Journal of Mining, Reclamation and Environment, 29*, 391–403.

Paredes, D., & Fleming-Munoz, D. (2021). Automation and robotics in mining: Jobs, income and inequality implications. *The Extractive Industries and Society, 8*, 189–193.

Parliamentary Advisory Council for Transport Safety (PACTS UK). (2022). Safe system. https://www.pacts.org.uk/safe-system (accessed April 20, 2022).

Pascoe, T. (2020). An evaluation of driverless haul truck incidents on a mine site: A mixed methodology. Unpublished PhD thesis. Curtin University of Technology.

Pascoe, T., McGough, S., & Jansz, J. (2022a). A multi-industry analysis of human-machine systems: The connection to truck automation. *World Safety Journal, XXXI*(1), 1–38.

Pascoe, T., McGough, S., & Jansz, J. (2022b). From truck driver awareness to obstacle detection: A tiger never changes its stripes. *World Safety Journal, XXXI*(2), 15–28.

Pascoe, T., McGough, S., & Jansz, J. (2022c). Haul truck automation: Beyond reductionism to avoid seeing turtles as rifles. *World Safety Journal, XXXI*(3), 26–38.

Pascoe, T., McGough, S., & Jansz, J. (2022d). The experiences of mineworkers interacting with driverless trucks: Risks, trust and teamwork. *World Safety Journal, XXXI*(4), 19–40.

Peng, S.S., Du, F., Cheng, J., & Li, Y. (2019). Automation in U.S. longwall coal mining: A state-of-the-art review. *International Journal of Mining Science and Technology, 29*, 151–159.

Perry, A. (2022). BHP is accelerating automation for safer operations. *Mining Safe to Work*, October 6. https://safetowork.com.au/bhp-is-accelerating-automation-for-safer-operations/ (accessed May 24, 2023).

Price, R., Cornelius, M., Burnside, L., & Miller, B. (2019). Mine planning and selection of autonomous trucks. (pp. 203-212) In E. Topal (Ed.), *Proceedings of the 28th International Symposium on Mine Planning and Equipment Selection*. New York: Springer.

Quinteiro, C., Quinteiro, M., & Hedstrom, O. (2001). Underground iron ore mining at LKAB, Sweden. In W.A. Hustrulid & R.L. Bullock (Eds.), *Underground Mining Methods: Engineering Fundamentals and International Case Studies* (pp. 361–368). Littleton: Society for Mining, Metallurgy, and Exploration, Inc.

Ralston, J.C., Hargrave, C.O., & Dunn, M.T. (2017). Longwall automation: Trends, challenges and opportunities. *International Journal of Mining Science and Technology, 27*, 733–739.

Ralston, J.C., Hargrave, C.O., & Hainsworth, D.W. (2006). Development of an autonomous conveyor-bolting machine for the underground coal mining industry. In S. Yuta, H. Asama, E. Prassler, T. Tsubouchi, & S. Thrun (Eds.) *Field and Service Robotics*. Springer Tracts in Advanced Robotics, vol. 24. Berlin, Heidelberg: Springer.

Ralston, J.C., Reid, D.C., Dunn, M.T., & Hainsworth, D.W. (2015). Longwall automation: Delivering enabling technology to achieve safer and more productive underground mining. *International Journal of Mining Science and Technology, 25*, 865–876.

Ralston, J.C., Reid, D.C., Hargrave, C., & Hainsworth, D. (2014). Sensing for advancing mining automation capability: A review of underground automation technology development. *International Journal of Mining Science and Technology, 24*, 305–310.

Redwood, N. (2023). Autonomous swarm haulage: The economics of autonomous haulage with small trucks. https://20485703.fs1.hubspotusercontent-na1.net/hubfs/20485703/Pronto-Whittle%20Swarm%20Haulage%20Study.pdf (accessed Aug 4, 2023).

Reid, D.C., Ralston, J.C., Dunn, M.T., & Hargrave, C.O. (2011). A major step forward in continuous miner automation. 11th Underground Coal Operators' Conference, pp. 165–171.

Rio Tinto (2019). How did one of the world's largest robots end up here? https://www.riotinto.com/en/news/stories/how-did-worlds-biggest-robot (accessed June 7, 2023).

Roberts, J.M., Duff, E.S., Corke, P.I., Sikka, O., Winstanley, G.J., & Cunningham, J. (2000). Autonomous control of underground mining vehicles using reactive navigation. Proceedings of the 2000 IEEE International Conference on Robotics & Automation, April, San Francisco, CA. pp. 3790–3795.

Scheding, S., Dissanayake, G., Nebot, E.M., & Durrant-Whyte, H. (1999) An experiment in autonomous navigation of an underground mining vehicle. *IEEE Transactions on Robotics and Automation, 15*, pp. 85–95.

Schiffbauer, W.H. (1997). *Accurate Navigation and Control of Continuous Mining Machines for Coal Mining*. Report of Investigations 9642. Pittsburgh PA: National Institute for Occupational Safety and Health .

Scott, I. A., Sullivan, C., & Staib, A. (2018). Going digital: A checklist in preparing for hospital-wide electronic medical record implementation and digital transformation. *Australian Health Review, 43*(3), 302–313.

Swart, C., Miller, F., Corbeil, P.A., Falmagne, V., & St-Arnaud, Luc. (2002). Vehicle automation in production environments. *Journal of the South African Institute of Mining and Metallurgy, 102,* 139–144.

Stentz, A., Bares, J., Singh, S., & Rowe, P. (1999). A robotic excavator for autonomous truck loading. *Autonomous Robots, 7,* 175–186.

Stringfellow, M. V. (2010). Accident analysis and hazard analysis for human and organizational factors. (Doctor of Philosophy). Massachusetts Institute of Technology, Massachusetts.

Tampier, C. Mascaro, M., & Ruiz-del-Solar, J. (2021). Autonomous loading system for Load-Haul-Dump (LHD) machines used in underground mining. *Applied Sciences, 11*(18), 8718.

Tariq, M., Gustafson, A., & Schunnesson, H. (2023) Training of load haul dump (LHD) machine operators: A case study at LKAB's Kiirunavaara mine, *Mining Technology.* https://doi.org/10.1080/25726668.2023.2217669

Taylor, B., & Spångberg, C. (2023). Shifting the needle on safety at the face - wireless development charging with Orica and Epiroc's Avatel™ charging solution. Proceedings of the 26th World Mining Congress. Brisbane, June, pp. 1856–1865.

Theiss (2021). Thiess applies SATS technology at Lake Vermont. https://thiess.com/news/innovation/thiess-applies-sats-technology-at-lake-vermont (accessed June 7, 2023).

Thompson, B. (2014). Safety trends in mining. Proceedings of the 6th International Platinum Conference. The Southern African Institute of Mining and Metallurgy.

Thompson, R.J. (2011). Mine road design and management in autonomous hauling operations: A research roadmap. Proceedings of the Second International Future Mining Conference. Paper 21.

Thompson, R.J. (2023). Hecla greens creek case study. https://www.cdc.gov/niosh/mining/features 2023automationpartnershipmeeting.html

UK Defence Standardization (2015). *Def Stan 00-251. Human Factors Integration for Defence Systems.*

UK Ministry of Defence (2021). *Joint Service Publication 912. Human Factors Integration for Defence Systems.*

University of Queensland. (2022). *Interviews with Glencore Champion 16 June and 23 August 2022.*

USA Department of Defense (2020). *Instruction 5000.88 Engineering Defense Systems*

USA Department of Defence (2022). *Directive 5000.01 The Defense Acquisition System.*

USA Federal Aviation Administration (1993). Order 9550.8. *Human Factors Policy.*

USA Federal Aviation Administration (2016). HF-STD-004a. *Human Factors Engineering Requirements.* https://hf.tc.faa.gov/publications/2016-09-human-factors-engineering-requirements/HF-STD-004A.pdf

USA National Aeronautics and Space Agency (2021). *Human Systems Integration Handbook.*

Valivaara, J. (2016). Automated drilling features for improving productivity and reducing costs in underground development. Proceedings of the Seventh International Conference and Exhibition on Mass Mining (MassMin 2016). AusIMM, Melbourne, pp. 723–730.

Vega, H., & Castro, R. (2020). Semi autonomous LHD operational philosophy for panel caving applications. In R. Castro, F. Báez, & K. Suzuki (Eds.), *MassMin 2020: Proceedings of the Eighth International Conference & Exhibition on Mass Mining* (pp. 1313–1321). Santiago: University of Chile.

Vicente, K. J. (1999). Cognitive work analysis: toward safe, productive, and healthy computer-based work. Mahwah, N.J.: Lawrence Erlbaum Associates.

Wang, J., & Huang, Z. (2017). The recent technological development of intelligent mining in Chine. *Engineering, 3,* 439–444.

WesTrac (2023). The evolution of automation in mining. https://www.westrac.com.au/resources/the-evolution-of-automation-in-australian-mining

Winstanley, G., Usher, K., Corke, P., Dunbabin, M., & Roberts, J. (2007). Dragline automation—A decade of development. *IEE Robotics and Automation Magazine*, September, 52–64.

Wolff Mining. Our services. https://www.wolffmining.com.au/services/automated-equipment/ (accessed November 22, 2022)

Xie, J., Yang, Z., Wang, X., & Wang, Y. (2018). A remote VR operation system for a fully mechanised coal-mining face using real-time data and collaborative network technology. *Mining Technology, 127*, 230–240.

Yaghini, A., Hall, R., & Apel, D. (2022). Autonomous and operator-assisted electric rope shovel performance study. *Mining, 2*, 699–711.

Index